ENGINEERING GENETIC CIRCUITS

CHAPMAN & HALL/CRC
Mathematical and Computational Biology Series

Aims and scope:

This series aims to capture new developments and summarize what is known over the whole spectrum of mathematical and computational biology and medicine. It seeks to encourage the integration of mathematical, statistical and computational methods into biology by publishing a broad range of textbooks, reference works and handbooks. The titles included in the series are meant to appeal to students, researchers and professionals in the mathematical, statistical and computational sciences, fundamental biology and bioengineering, as well as interdisciplinary researchers involved in the field. The inclusion of concrete examples and applications, and programming techniques and examples, is highly encouraged.

Series Editors

Alison M. Etheridge
Department of Statistics
University of Oxford

Louis J. Gross
Department of Ecology and Evolutionary Biology
University of Tennessee

Suzanne Lenhart
Department of Mathematics
University of Tennessee

Philip K. Maini
Mathematical Institute
University of Oxford

Shoba Ranganathan
Research Institute of Biotechnology
Macquarie University

Hershel M. Safer
Weizmann Institute of Science
Bioinformatics & Bio Computing

Eberhard O. Voit
The Wallace H. Couter Department of Biomedical Engineering
Georgia Tech and Emory University

Proposals for the series should be submitted to one of the series editors above or directly to:
CRC Press, Taylor & Francis Group
4th, Floor, Albert House
1-4 Singer Street
London EC2A 4BQ
UK

Published Titles

Bioinformatics: A Practical Approach
Shui Qing Ye

Cancer Modelling and Simulation
Luigi Preziosi

Combinatorial Pattern Matching Algorithms in Computational Biology Using Perl and R
Gabriel Valiente

Computational Biology: A Statistical Mechanics Perspective
Ralf Blossey

Computational Neuroscience: A Comprehensive Approach
Jianfeng Feng

Data Analysis Tools for DNA Microarrays
Sorin Draghici

Differential Equations and Mathematical Biology
D.S. Jones and B.D. Sleeman

Engineering Genetic Circuits
Chris J. Myers

Exactly Solvable Models of Biological Invasion
Sergei V. Petrovskii and Bai-Lian Li

Gene Expression Studies Using Affymetrix Microarrays
Hinrich Göhlmann and Willem Talloen

Handbook of Hidden Markov Models in Bioinformatics
Martin Gollery

Introduction to Bioinformatics
Anna Tramontano

An Introduction to Systems Biology: Design Principles of Biological Circuits
Uri Alon

Kinetic Modelling in Systems Biology
Oleg Demin and Igor Goryanin

Knowledge Discovery in Proteomics
Igor Jurisica and Dennis Wigle

Meta-analysis and Combining Information in Genetics and Genomics
Rudy Guerra and Darlene R. Goldstein

Modeling and Simulation of Capsules and Biological Cells
C. Pozrikidis

Niche Modeling: Predictions from Statistical Distributions
David Stockwell

Normal Mode Analysis: Theory and Applications to Biological and Chemical Systems
Qiang Cui and Ivet Bahar

Optimal Control Applied to Biological Models
Suzanne Lenhart and John T. Workman

Pattern Discovery in Bioinformatics: Theory & Algorithms
Laxmi Parida

Spatial Ecology
Stephen Cantrell, Chris Cosner, and Shigui Ruan

Spatiotemporal Patterns in Ecology and Epidemiology: Theory, Models, and Simulation
Horst Malchow, Sergei V. Petrovskii, and Ezio Venturino

Stochastic Modelling for Systems Biology
Darren J. Wilkinson

Structural Bioinformatics: An Algorithmic Approach
Forbes J. Burkowski

The Ten Most Wanted Solutions in Protein Bioinformatics
Anna Tramontano

Chapman & Hall/CRC Mathematical and Computational Biology Series

ENGINEERING GENETIC CIRCUITS

CHRIS J. MYERS

University of Utah
Salt Lake City, Utah, U. S. A.

CRC Press
Taylor & Francis Group
Boca Raton London New York

CRC Press is an imprint of the
Taylor & Francis Group, an **informa** business

A CHAPMAN & HALL BOOK

CRC Press
Taylor & Francis Group
6000 Broken Sound Parkway NW, Suite 300
Boca Raton, FL 33487-2742

First issued in paperback 2018

© 2010 by Taylor and Francis Group, LLC
CRC Press is an imprint of Taylor & Francis Group, an Informa business

No claim to original U.S. Government works

ISBN-13: 978-1-4200-8324-8 (hbk)
ISBN-13: 978-1-138-37273-3 (pbk)

Library of Congress Cataloging-in-Publication Data

Myers, Chris J., 1969-
    Engineering genetic circuits / author, Chris J. Myers.
        p. cm. -- (Chapman & Hall/CRC mathematical and computational biology series)
    Includes bibliographical references and index.
    ISBN 978-1-4200-8324-8 (hard back : alk. paper)
    1. Logic circuits. 2. Combinatorial circuits. 3. Molecular computers. 4. Molecular electronics. 5. Genetic engineering. I. Title.

TK7868.L6M95 2010
570.285--dc22   2009019959

Visit the Taylor & Francis Web site at
http://www.taylorandfrancis.com

and the CRC Press Web site at
http://www.crcpress.com

To Ching and John

Contents

List of Figures   xiii

List of Tables   xvii

Foreword   xix

Preface   xxiii

Acknowledgments   xxvii

1 An Engineer's Guide to Genetic Circuits   **1**
- 1.1 Chemical Reactions 1
- 1.2 Macromolecules 4
- 1.3 Genomes 8
- 1.4 Cells and Their Structure 9
- 1.5 Genetic Circuits 13
- 1.6 Viruses 16
- 1.7 Phage λ: A Simple Genetic Circuit 17
  - 1.7.1 A Genetic Switch 19
  - 1.7.2 Recognition of Operators and Promoters 26
  - 1.7.3 The Complete Circuit 29
  - 1.7.4 Genetic Circuit Models 35
  - 1.7.5 Why Study Phage λ? 36
- 1.8 Sources 39
- Problems 40
- Appendix 43

2 Learning Models   **51**
- 2.1 Experimental Methods 52
- 2.2 Experimental Data 57
- 2.3 Cluster Analysis 59
- 2.4 Learning Bayesian Networks 62
- 2.5 Learning Causal Networks 68
- 2.6 Experimental Design 79
- 2.7 Sources 80
- Problems 81

3   Differential Equation Analysis   **85**
   3.1   A Classical Chemical Kinetic Model   86
   3.2   Differential Equation Simulation   88
   3.3   Qualitative ODE Analysis   92
   3.4   Spatial Methods   97
   3.5   Sources   98
        Problems   99

4   Stochastic Analysis   **103**
   4.1   A Stochastic Chemical Kinetic Model   104
   4.2   The Chemical Master Equation   106
   4.3   Gillespie's Stochastic Simulation Algorithm   107
   4.4   Gibson/Bruck's Next Reaction Method   111
   4.5   Tau-Leaping   115
   4.6   Relationship to Reaction Rate Equations   117
   4.7   Stochastic Petri-Nets   119
   4.8   Phage λ Decision Circuit Example   120
   4.9   Spatial Gillespie   124
   4.10 Sources   127
        Problems   127

5   Reaction-Based Abstraction   **131**
   5.1   Irrelevant Node Elimination   132
   5.2   Enzymatic Approximations   133
   5.3   Operator Site Reduction   136
   5.4   Statistical Thermodynamical Model   145
   5.5   Dimerization Reduction   151
   5.6   Phage λ Decision Circuit Example   153
   5.7   Stoichiometry Amplification   154
   5.8   Sources   154
        Problems   157

6   Logical Abstraction   **161**
   6.1   Logical Encoding   163
   6.2   Piecewise Models   165
   6.3   Stochastic Finite-State Machines   170
   6.4   Markov Chain Analysis   174
   6.5   Qualitative Logical Models   180
   6.6   Sources   184
        Problems   185

7 Genetic Circuit Design   187
   7.1 Assembly of Genetic Circuits 188
   7.2 Combinational Logic Gates 189
   7.3 PoPS Gates 198
   7.4 Sequential Logic Circuits 198
   7.5 Future Challenges 211
   7.6 Sources 212
       Problems 213

Solutions to Selected Problems   215

References   237

Glossary   247

Index   273

List of Figures

0.1 Engineering approach xxvi

1.1 Nucleic acids 6
1.2 Protein structure 7
1.3 Introns and exons 10
1.4 Prokaryotes and eukaryotes 11
1.5 An overview of transcription and translation 14
1.6 Transcription and translation 14
1.7 Types of viruses 16
1.8 Bacteriophage Lambda (Phage λ) 18
1.9 Phage λ developmental pathways 18
1.10 The O_R operator 19
1.11 Binding of CI dimers to O_R operator sites 19
1.12 Binding of Cro dimers to O_R operator sites 20
1.13 The effect of CI and Cro bound to each operator site 21
1.14 Position of CI_2 bound to O_R versus CI_2 concentration 22
1.15 Position of Cro_2 bound to O_R versus Cro_2 concentration . . . 23
1.16 Cooperativity of CI_2 binding 24
1.17 Effect of cooperativity of CI_2 molecules 25
1.18 Induction 26
1.19 Amino acid-base pair interactions 28
1.20 The λ genome 29
1.21 Patterns of gene expression 31
1.22 The action of N 32
1.23 Retroregulation of Int 33
1.24 Integration and induction 34
1.25 A portion of the phage λ decision circuit 35
1.26 Degradation reactions 37
1.27 Open complex formation reactions 37
1.28 Dimerization reactions 37
1.29 Repression reactions 37
1.30 Activation reactions 38
1.31 Complete model for the CI/CII portion of phage λ 38
1.32 Phage λ decision circuit 43
1.33 Chemical reaction network model of the promoter P_{RE} 44
1.34 Model with one species bound to the O_R operator 45
1.35 Model with two species bound to the O_R operator 46

1.36 Model with three species bound to the O_R operator 47
1.37 Model for CI and Cro dimerization and degradation 47
1.38 Model for N production and degradation 48
1.39 Model for CIII production 48
1.40 Model for CII production 49
1.41 Model for CII and CIII degradation 49

2.1   Fluorescent proteins 52
2.2   Procedure for a DNA microarray experiment 53
2.3   2D gel electrophoresis and mass spectrometry 55
2.4   Method for two-hybrid screening 56
2.5   Time series experimental data 58
2.6   K-means example 60
2.7   Clustering examples of microarray gene expression data . . . 61
2.8   Bayesian network examples 63
2.9   A Bayesian model for the phage λ decision circuit 66
2.10 Another Bayesian model for the phage λ circuit 67
2.11 A DBN for the phage λ decision circuit 68
2.12 Possible models for the phage λ decision circuit 69
2.13 N's probability of increasing with $i = \langle n, r, r, n, n \rangle$ 72
2.14 Thresholds for determining votes 73
2.15 N's probability of increasing with $i = \langle n, r, r, n, n \rangle, G = \{N\}$. 74
2.16 CIII's probability of increasing with $i = \langle r, n, n, r, n \rangle$ 77
2.17 CIII's probability of increasing with $i = \langle n, n, n, n, r \rangle$ 78
2.18 CIII's probability of increasing with $i = \langle r, n, n, r, n \rangle$ 78
2.19 The final genetic circuit learned for the phage λ circuit 79
2.20 Alternative models for experimental design example 80

3.1   ODE model example 87
3.2   Euler methods 89
3.3   Midpoint method 90
3.4   ODE simulation of CI and CII 91
3.5   Saddle-node bifurcation example 93
3.6   Transcritical bifurcation example. 94
3.7   Pitchfork bifurcation example. 95
3.8   Nullcline for the CI/CII portion of phage λ 96
3.9   Examples of spatial configurations 97

4.1   Gillespie's stochastic simulation algorithm (SSA) 109
4.2   Comparison of SSA to ODE ($P_R = P_{RE} = 1$, RNAP= 30) . . 110
4.3   Comparison of SSA to ODE ($P_R = P_{RE} = 10$, RNAP= 300) . 111
4.4   Gillespie's first reaction method 112
4.5   Gibson and Bruck's next reaction method 113
4.6   Dependency graph for the CI/CII portion of phage λ 114
4.7   Indexed priority queue example 115

4.8   Explicit tau-leaping simulation algorithm 117

4.9   A simple SPN 120

4.10 An SPN for part of P_{RE} with CI dimerization/degradation . 121

4.11 Time courses for CI_2 and Cro_2 123

4.12 Probability of lysogeny versus MOI 124

4.13 Lysogenic fraction versus API 125

4.14 Lysogenic fraction versus API for O-T- and O-N- mutants . . 125

4.15 Discrete subvolumes used by the spatial Gillespie method . . 126

5.1   Irrelevant node elimination 132

5.2   Enzymatic approximations 134

5.3   Competitive enzymatic reaction example 137

5.4   Fractions of operator sites free of repressor 139

5.5   Fractions of operator sites bound to an activator 140

5.6   Operator site reduction 143

5.7   Similar reaction and modifier constant reductions 144

5.8   Configurations of CI_2 molecules on the O_R operator 146

5.9   Probability of repression of the P_R and P_{RM} promoters . . . 148

5.10 Open complex rate for P_R and P_{RM} 149

5.11 Dimerization reduction 152

5.12 Abstracted reaction-based model for CI/CII 153

5.13 Abstracted reaction-based model for phage λ 155

5.14 Stoichiometry amplification example 156

6.1   Example electrical circuits 162

6.2   All reactions involved in $E.$ $coli$ metabolism 162

6.3   Activity of the P_{RM} promoter 164

6.4   Example regulatory domains 165

6.5   Flow graph for the PLDE model of the CI/CII example . . . 168

6.6   LHPN model for the CI/CII example 169

6.7   Piecewise model for the phage λ decision circuit 169

6.8   Reaction splitization 171

6.9   CTMC example 181

6.10 Comparison of CTMC results to experimental data 182

6.11 State graph examples 183

7.1   BioBricks^{TM} 188

7.2   Genetic inverter 189

7.3   Genetic oscillator 190

7.4   Genetic NAND gate 191

7.5   Genetic NOR gate 192

7.6   Genetic AND gate using a NAND gate and inverter 193

7.7   Genetic OR gate using two inverters and a NAND gate . . . 194

7.8   Genetic AND gate using chemical inducers 195

7.9   Genetic OR gate using chemical inducers 196

7.10 Genetic AND gate using one gene 197
7.11 PoPS examples 199
7.12 Genetic AND gate with memory 200
7.13 Genetic toggle switch 201
7.14 Genetic toggle Muller C-element 203
7.15 Nullcline analysis for the genetic toggle C-element 204
7.16 Stochastic analysis for the genetic C-elements 206
7.17 Genetic majority Muller C-element 207
7.18 Genetic speed-independent Muller C-element 208
7.19 Effects of repression strength and cooperativity on failure rate 209
7.20 Effects of decay rate on the toggle switch genetic C-element . 210

8.1   Chemical reaction-based model for Problem 1.1 216
8.2   A Bayesian network model for Problem 2.3 217
8.3   Causal network for Problem 2.3.5 218
8.4   ODE model for Problem 3.4 220
8.5   Simulation results for Problem 3.8 221
8.6   Simulation results for Problem 4.3 223
8.7   SPN for the reaction-based model from Problem 1.3 224
8.8   Abstracted reaction-based model for Problem 5.7 229
8.9   CTMC for Problem 6.1.4 232
8.10 State graph for Problem 6.1.5 233
8.11 A genetic XOR gate 234
8.12 A genetic PoPS XOR gate 235
8.13 A genetic toggle 236

List of Tables

1.1 The genetic code 7
1.2 Near symmetry in the operator sequences 27
1.3 λ promoters 28

2.1 Synthetic gene expression data for phage λ 65
2.2 Probabilities, ratios, and votes for CII repressing CIII 71
2.3 Probabilities, ratios, and votes for $\langle n, r, r, n, n \rangle$ with $G = \{N\}$ 75
2.4 Probabilities and ratios found by the Score function 75
2.5 Votes and scores found by the Score function 76
2.6 Votes and scores calculated when combining influence vectors 76
2.7 Scores found when competing potential influence vectors . . . 77

5.1 Free energies for O_R configurations with CI_2 146
5.2 Values for $[CI_2]$ for half occupation (units of 3nM) 147
5.3 Resolved interaction free energies for O_R 148
5.4 Free energies for O_R configurations 150

Foreword

This is a book about conversion and expression. It is a clearly written textbook that teaches the reader how to convert biomolecular observations into mathematical models of genetic circuits. The author also shows how to use such models to discover the principles of circuit operation with suitable resolution to predict, control, and design function. These circuits, encoded in the one-dimensional information of the genome sequence, are expressed by the cellular machinery into the teeming variety of three-dimensional life with which we are familiar. Aside from evolution itself, I can think of no more awe-inspiring transformation.

Evolution implicitly plays its part in this book. The fact that there are generalizations and abstractions to be made about the architecture and organization of the cellular machinery is traceable to the mechanisms of evolution. From the origin of life approximately 3.5 billion years ago, via the extraordinary mechanisms of mutation and genetic transfer, the genome has explored a vast space of possible configurations of cellular networks that have been subject to natural selection. "Designs" that lead to efficient procreation survive with the rest eventually weeded out. There is a good argument that natural selection and the mechanisms of mutation incidentally leads to evolvability—the ability of a design to withstand and be quantitatively tuned by small variation and, in addition, have significant numbers of routes to reconfiguration for new, plausibly useful, function. Evolvability itself may lead to the proliferation and reuse of "motifs" of function ranging from macromolecular signal sequences and domains to whole subnetwork architectures. That evolution may drive towards recurrent molecular "designs" and circuit motifs is one of the key observations underlying the promise and power of Systems Biology. To those who, by profession, appreciate and attempt to make sense of the function of complex systems, life and the biological circuits that underlie it provide a nearly irresistible allure. It is increasingly clear that there are principles of organization, architecture, and (evolutionary) design in biological circuits, but there are many details and deep principles yet to be discovered.

This book is also about the conversion and expression of the author, who is a world-class engineer and has also written an excellent textbook on Asynchronous Circuit Design for electronics. I am not an electrical engineer, but as I understand it, engineering with a clock, enforcing synchronous operation of a circuit, greatly simplifies the design of circuits since it is predictable when (and where) signals are coming from. Thus, the timing of arrivals of signals is, at most, a second order consideration. However, as such circuits increase in complexity and size, it becomes impractical to ensure that even the signal from the clock itself is received simultaneously by all elements. The assumptions of synchrony breakdown leading to immense complication in design. Professor Myers is an expert in controlling this complication. Biological circuits do not really have a central clock. Worse, they do not adhere to homogeneous Boolean abstractions nor are all the inputs, outputs, and internal mechanisms unambiguously known even for relatively small subsystems of cellular function. One can see how an engineer of Professor Myers' pedigree might be drawn to this different and difficult type of circuitry.

Professor Myers' book is really about Systems Biology. While the origins of the term are a bit cloudy, it is fairly clear that it arose as an identifiable discipline in the mid-sixties though the intellectual foundations I would argue came quite a bit earlier. But the 1960's is when both the genetic code was cracked and the molecular basis for the observation that the expression of genes and the activities of their encoded proteins could be regulated by their own and other genes products was discovered. This "Central Dogma" was solidified—DNA is transcribed into RNA and RNA is translated into protein, and it was clear these macromolecules formed complex dynamic networks of interaction. It became equally clear that the study of how these lines of connection logically translated nearly static genotype to dynamic phenotype was a large and sophisticated study in itself. This kind of approach is in contrast to the powerful and prerequisite discovery science that is still the dominant paradigm in biology. Once the parts of the system are known and the mechanisms of their individual interactions are characterized, there is still the problem of understanding how that system, so exquisitely described, functions.

Imagine that I'm handing you the schematic for one of Professor Myers' asynchronous digital circuits complete with a description of how every component worked. How would you determine what it does and how it encodes and transforms its signals into dynamical calculations? What are the key parameters that control this function, and whose variation might lead to failure or change of behavior? How would you understand the principles of its function? (After all, there are many circuits that might implement a given function—how does this one work and why was this design selected?) How would you uncover the structure of its design? Is the system "modular" in expression or control? Are there subcircuits that themselves have useful function? Are there elements and architectures that confer robustness to uncertainty in the environment? How can we perturb the circuit to move it from a possibly unknown state to a desired one? For biological circuits, there are also questions that skate dangerously close to teleology (I am a not-so-closeted teleologist myself). Is the circuit optimal for something identifiably "engineering"? By its existence, a biological system must be very good at surviving and procreating, but how? Is energy usage or temporal response optimal in some way? Is the particular architecture necessary and/or sufficient for a particular function such as, for example, gradient sensing? Does it implement a winning strategy in a discoverable evolutionary game? Is the architecture especially amenable to "upgrade" or use of pre-existing parts from other devices? Or is it an accident of largely vertical inheritance from progenitors that arose under different conditions?

I believe these questions, applied to biological circuits, are the particular but non-exclusive bailiwick of Systems Biology. The specific experimental tools, theoretical frameworks, and computational algorithms that apply to these problems are changeable and distinct. The choice of tools, of course, depends on the resolution or scale of the question under consideration—and technology in this area is changing incredibly quickly. It is not the tools but the questions about the integrated function of cellular networks that define Systems Biology in my mind. If molecular or mechanistic discovery comes along for a ride, that is a welcome bonus. That the focus is on circuit behavior and (evolutionary) design is what so draws engineers like Professor Myers to the field.

When Chris and I met at Stanford in the mid-nineties, I was working to understand how the physical mechanisms of the cell might have evolved to implement comprehensible "engineering" function, and how such function might be inferred from data. Chris was an engineering student working on systems complex enough to require breaking the standard electronic engineering paradigm and developing new principles of circuit design. With our mutual friend, Michael Samoilov, we began an exchange that continues today, learning from each other and sharing our appreciation for the designs of Dawkins old Blind Watchmaker, evolution.

I will claim credit for introducing Professor Myers to the venerable and very fascinating Phage Lambda as the model system of choice to test nearly all ideas about Systems Biology. Phage Lambda was involved in an astonishing number of fundamental discoveries in genetics, biochemistry, virology, and development. Lambda has stood, since the inception of its study mid-last century, as one of the few biological systems that has strongly linked physical, engineering, and biological scientists. The impact of Lambda and its mechanisms on biology and biotechnology would be hard to overstate. At its heart are two famous decisions: the decision upon infection to either create many progeny, thereby killing the host (lysis), or to integrate into the host chromosome (lysogeny), thereby conferring immunity to further infection; and the decision, after integration, to re-enter the lytic path if host conditions become unfavorable. Birth, death, predation, temporary symbiosis, and decision-making are all encoded in one 50 kb genome package (ignoring host functions). Together with my colleagues, Harley McAdams and John Ross, I was deeply immersed in models of its networks to study the implications of molecular discreteness and noise in the robustness of cellular decisions. For an engineer like Chris, the famous "switch" was a clear entry place and provided a common point of reference for our discussions. Chris took off on his own, deepening his understanding of Systems Biology through research on this model system. It is gratifying to see Lambda as an organizing example used so effectively in this text.

One of the defining properties of the study of Systems Biology—even for this relatively simple system—is that no one can agree on what one needs to know to study it effectively. Even more than half a century after its discovery, there are still new mechanisms being discovered about the functioning of the Lambda switch. Thus, the classical biologists demand (rightly so) that anyone studying the switch must be an expert in biology and understand Lambda lore, in particular, quite well because otherwise any models of its architecture and function will not properly capture the mechanistic complexity or uncertainty about its operation. The biochemists and biophysicists will insist that the particulars of the mechanisms—how exactly the proteins interact and degrade, how DNA loops to form a key component of the switch, how the complex thermodynamics of multicomponent promoter binding leads to the proper ordering of states for stabilizing the two decisions, how the stochastic effects due to the discrete nature of the chemistry leads to diversity in behavior—all must be considered carefully or formally discounted to justify explanations for how the system works. The computational biologists and systems biologists require, on top of the above, deep understanding of biological data analysis, numerical algorithms, graph theory, and dynamical systems and control theory in order to build appropriate models justified by data and to analyze them properly, and a fine sense of how to interpret and abstract the principles of control buried in the dunes of detail about the system.

Students, throw up your hands! How do you start with all this deep and complex science and engineering to learn? As a first step, we must learn to live with our ignorance, and only then can we begin to fight to eradicate it. The frontier of science is where ignorance begins, and for Systems Biology, there is plenty of undiscovered land. This book by Professor Myers is one of the few texts in the area that gently brings the uninitiated to these edges. I congratulate him for his achievement—*Engineering Genetic Circuits* admirably touches on much of the "required" knowledge above while creating a minimal toolset with which beginning students can confidently venture into this exciting new territory of Systems Biology.

Adam Arkin
Berkeley, California

Preface

Engineering is the art or science of making practical.

—Samuel C. Florman

One man's "magic" is another man's engineering.

—Robert Heinlein

In the 21st century, biology is being transformed from a purely lab-based science to also an information science. As such, biologists have had to draw assistance from those in mathematics, computer science, and engineering. The result has been the development of the fields of *bioinformatics* and *computational biology* (terms often used interchangeably). The major goal of these fields is to extract new biological insights from the large and noisy sets of data being generated by high-throughput technologies. Initially, the main problems in bioinformatics were how to create and maintain databases for massive amounts of DNA sequence data. Addressing these challenges also involved the development of efficient interfaces for researchers to access, submit, and revise data. Bioinformatics has expanded into the development of software that also analyzes and interprets these data. While bioinformatics has come to mean many things, this text uses the term bioinformatics to refer to the analysis of *static* data such as DNA and protein sequences, techniques for finding genes or evolutionary patterns, and *cluster analysis* of microarray data. Algorithms for bioinformatics are not covered in this text.

The focus of this text is modeling, analysis, and design methods for *systems biology*. Systems biology has been the subject of several books (Kitano, 2001; Alberghina and Westerhoff, 2005; Palsson, 2006; Wilkinson, 2006; Alon, 2007; Konopka, 2007), each of which give it a somewhat different meaning. This book uses the term to mean the study of the mechanisms underlying complex molecular processes as integrated into systems or pathways made up of many interacting genes and proteins. In other words, systems biology is concerned with the analysis of *dynamic* models. While it has long been known that developing dynamic models of complete systems is essential to understanding biological processes, it is only recently that the emergence of new high-throughput experimental data acquisition methods has made it possible to explore such computational models. Some example experimental techniques include *cDNA microarrays* and *oligonucleotide chips* (Brown and Botstein, 1999; Lipschutz *et al.*, 1999; Lockhart and Winzeler, 2000; Baldi and Hatfield, 2002), *mass spectrometric identification of gel-separated proteins* (Kahn, 1995; Mann, 1999; Pandey and Mann, 2000; Zhu and Snyder, 2001), *yeast two-hybrid systems* (Chien *et al.*, 1991), and *genome-wide location analysis* (ChiP-to-chip) (Ren and Dynlacht, 2004).

Systems biology involves the collection of large experimental data sets using high-throughput technologies, the development of mathematical models that predict important elements in this data, design of software to accurately and efficiently make predictions *in silico* (i.e., on a computer), quality assessment of models by comparing numerical simulations with experimental data, and the design of new synthetic biological systems. The ultimate goal of systems biology is to develop models and analytical methods that provide reasonable predictions of experimental results. While it will never replace experimental methods, the application of computational approaches to gain understanding of various biological processes has the promise of helping experimentalists work more efficiently. These methods also may help gain insight into biological mechanisms, information which may not be obtained from any known experimental methods. Eventually, it may be possible that such models and analytical techniques could have substantial impact on our society such as aiding in drug discovery.

Systems biologists analyze several types of molecular systems, including *genetic regulatory networks*, *metabolic networks*, and *protein networks*. The primary focus of this book is genetic regulatory networks referred to as *genetic circuits* in this book. These circuits regulate gene expression at many molecular levels through numerous feedback mechanisms. Chapter 1 presents the basic molecular biology and biochemistry principles that are needed to understand these circuits. A few bacterial genetic circuits are well understood. One such circuit is the lysis/lysogeny decision circuit of the phage λ. This circuit is described in Chapter 1 and used as a running example throughout this book.

During the genomic age, standards for representing sequence data were (and still are) essential. Data collected from a variety of sources could not be easily used by multiple researchers without a standard data format. For systems biology, standard data formats are also being developed. One format that seems to be gaining some traction is the *systems biology markup language* (SBML) (see `http://sbml.org`) (Hucka *et al.*, 2003), which is an XML-based language for representing chemical reaction networks. All the types of models described in this book can be reduced to a set of bio-chemical reactions. The basic structure of an SBML model is a list of *chemical species* coupled with a list of *chemical reactions*. Each chemical reaction includes a list of *reactants*, *products*, and *modifiers*. It also includes a mathematical description of the *kinetic rate law* governing the dynamics of this reaction. SBML is not a language for use in constructing models by hand. Fortunately, several *graphical user interfaces* (GUIs) have been developed for entering or drawing up chemical reaction networks, which then can be exported in the SBML format.

Another essential item in the genomic age was the development of *biological databases*, which provide repositories for storing large bodies of data that can be easily updated, queried, and retrieved. Databases have been developed to store different sets of information ranging from nucleotide sequences within GenBank (http://www.ncbi.nlm.nih.gov/GenBank/) to biomedical literature at PubMed (http://www.ncbi.nlm.nih.gov/pubmed/). Recently, there has also been developed a new database for SBML models of various biochemical systems (http://www.ebi.ac.uk/biomodels/).

A final essential piece of the puzzle is the development of tools for analysis. An excellent list of bioinformatics tools can be found on NCBI's website at http://www.ncbi.nlm.nih.gov/Tools/. A list of systems biology tools that support the SBML language can be found at the SBML website at http://sbml.org/SBML_Software_Guide. This text concentrates on describing the methods used by such tools.

Given our vast experience in reasoning about complex circuits and systems, engineers are uniquely equipped to assist with the development of tools for the modeling and analysis of genetic circuits. It has been shown that viewing a genetic circuit as an electronic circuit can yield new insights and open up the application of engineering tools to genetic circuits (McAdams and Shapiro, 1995). Therefore, as in the sequencing of the human genome, collaborations between engineers and biologists will be essential to the success of systems biology. The major goal of this textbook is to facilitate these collaborations.

An engineering approach involves three parts as shown in Figure 0.1. First, engineers examine experimental data in order learn mathematical models. Second, engineers develop efficient abstraction and simulation methods to analyze these models. Finally, engineers use these analytical methods to guide the design of new circuits. This book discusses all three aspects of this engineering approach as applied to genetic circuits. Chapter 2 describes modern experimental techniques and methods for learning genetic circuit models from the data generated by these experiments. The next four chapters explore methods for analyzing these genetic circuit models. Perhaps, the most common method for modeling and analysis uses differential equations, a topic which is the subject of Chapter 3. In genetic circuits, however, the numbers of molecules involved are typically very small, thereby requiring the use of stochastic analysis; these methods are described in Chapter 4. Stochastic analysis can be extremely complex limiting its application often to only the simplest systems. To reduce this complexity, Chapter 5 presents several reaction-based abstraction methods to simplify the models while maintaining reasonable accuracy. Since the state space of these models is still often quite large, Chapter 6 presents logical abstraction to reduce this state space and further improve analysis time. Finally, using these analytical methods, researchers are beginning to design synthetic genetic circuits as described in Chapter 7. It is our hope that this book will prove to be useful to both engineers who wish to learn about genetic circuits and to biologists who would like to learn about engineering techniques that can be used to study their systems of interest.

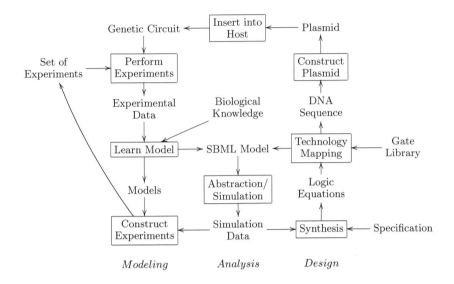

FIGURE 0.1: Engineering approach to modeling, analysis, and design of genetic circuits.

This textbook has been used several times in an advanced undergraduate/graduate level course on the modeling, analysis, and design of biological circuits. In a semester version of this course, students select their own biological circuit and learn about them from research papers. Homework each week includes a paper and pencil problem and a software tutorial using the phage λ example. Students then repeat the model design or analysis using their selected circuit. Finally, students complete a more extensive final project, which ranges from modeling and analysis to the development of software depending on the student's background and interests. Throughout, the students use our iBioSim tool, which supports most of the topics discussed in this textbook. This tool allows one to construct models, learn from experimental data, and perform either differential or stochastic analyses utilizing automatic reaction-based and logical abstractions. Course examples including lecture materials and assignments as well as our iBioSim software are freely available from:

http://www.async.ece.utah.edu/BioBook/

Chris J. Myers
Salt Lake City, Utah

Acknowledgments

This book would never have been started without the influence of Professor Adam Arkin and Dr. Michael Samoilov of the University of California at Berkeley. When Michael and I were graduate students at Stanford University, he introduced me to Adam, thus launching many discussions about the relationship between asynchronous electronic circuits and genetic circuits. The result was Adam joining my dissertation committee. In 2002, during my sabbatical at Stanford University, we revisited these conversations, thus triggering much of the research described in this textbook. I would also like to thank Professor David Dill of Stanford who hosted my sabbatical and used a draft of this text in one of his courses. Finally, I would like to thank Daniel Gillespie for several stimulating discussions.

Much of the work described in this textbook was conducted by my graduate students. Dr. Nathan Barker, now an Assistant Professor at Southern Utah University, developed the causal model learning methods described in Chapter 2. Dr. Hiroyuki Kuwahara, now a research associate at the Microsoft Research/University of Trento Centre for Systems and Computational Biology in Trento, Italy, developed most of the abstraction methods described in Chapters 5 and 6. Nam Nguyen, now a Ph.D. student at the University of Texas in Austin, designed the genetic Muller C-element presented in Chapter 7. All three of these graduate students as well as my current students, Curtis Madsen and Kevin Jones, have been involved in the development of iBioSim, the tool that implements many of the methods described in this textbook.

I would like to thank Luna Han, Marsha Pronin, Michele Dimont, and Shashi Kumar of Taylor and Francis Publishing Group for their assistance in getting this book ready for publication. I would also like to thank Baltazar Aguda of Ohio State University, Lingchong You of Duke University, and the other anonymous reviewers whose comments helped improve this book.

Finally, I would especially like to thank my family, Ching and John, for being patient with me during the writing process. Without their love and support, this book would not have been possible.

C.J.M.

Chapter 1

An Engineer's Guide to Genetic Circuits

Biology has at least 50 more interesting years.

—James D. Watson

DNA makes RNA, RNA makes protein, and proteins make us.

—Francis Crick

This chapter gives a brief introduction to the biology and biochemistry necessary to understand genetic circuits. Obviously, the material covered in this chapter is quite basic as it is usually the topic of whole courses. It should, however, give the grounding necessary to begin studying the modeling, analysis, and design of genetic circuits.

This chapter begins with a discussion of *chemical reactions*, the basic mechanism used by all life processes. Section 1.2 describes the basic building blocks of life, *macromolecules*. Section 1.3 describes *genomes* which store the instructions for all life's processes. Section 1.4 presents *cells* and their structure. Section 1.5 introduces the genetic circuits used by cells to regulate the production of *proteins* within cells. Section 1.6 describes *viruses*. Finally, Section 1.7 presents our running example, the genetic circuit from the phage λ virus that controls its choice of developmental pathway.

1.1 Chemical Reactions

Atoms are the basic building blocks for all matter whether living or not. Although there are more than a 100 different types of atoms, about 98 percent of the mass of any living organism is made up of only six types: hydrogen (H), carbon (C), nitrogen (N), oxygen (O), phosphorus (P), and sulfur (S). It is truly amazing what nature has constructed from such a simple set of basic building blocks.

All materials that make up a living organism are created or destroyed via chemical reactions. Chemical reactions combine atoms to form *molecules* and combine simpler molecules to form more complex ones. Atoms form molecules by binding together via *covalent, ionic*, and *hydrogen bonds*. Chemical reactions can also work in reverse to turn complex molecules into simpler molecules or atoms.

A simple chemical reaction for the formation of water from hydrogen and oxygen is shown below:

$$2H_2 + O_2 \xrightarrow{k} 2H_2O$$

where H_2, O_2, and H_2O are known as *chemical* (or *molecular*) *species*. In this equation, the subscripts in H_2 and O_2 indicate that the hydrogen and oxygen are present in *dimer* form (i.e., a molecule composed of two like atoms or molecules). The molecules H_2 and O_2 are known as the *reactants* for this reaction. The water molecule, H_2O, is composed of two hydrogen atoms and one oxygen atom, and it is known as the *product* for this reaction. The number 2 in front of H_2 and H_2O indicates that two hydrogen dimers are used in the reaction and two water molecules are produced. These numbers (along with the implicit one in front of O_2) are known as the *stoichiometry* of the reaction. Since matter must be conserved by chemical reactions, the numbers of each atom on both sides of the equation must be equal. Note that many chemical reactions shown in this book may not have this property when reactants or products are not listed to simplify the presentation.

The k above the arrow is known as the *rate constant* for this chemical reaction. This value indicates how likely or how fast this reaction typically occurs. The rate constant is used in many of the modeling techniques described in this book, but unfortunately, it is often difficult to determine accurately for bio-chemical reactions. The rate of a chemical reaction is governed by the *law of mass action*. Namely, the rate of a reaction is determined by this rate constant and the concentrations of the reactants raised to the power of their stoichiometry. Using the law of mass action for the reaction above, the rate of water formation via this reaction is:

$$\frac{d[H_2O]}{dt} = 2k[H_2]^2[O_2]$$

where $[H_2O]$, $[H_2]$, and $[O_2]$ represent the concentration of water, hydrogen dimers, and oxygen dimers, respectively. The 2 in front of the k signifies that each reaction produces two molecules of water.

Chemical reactions must obey the *laws of thermodynamics*. The first law states that energy can be neither created nor destroyed, and the second law states that *entropy* (i.e., the disorder in the universe) must increase or stay the same. These two laws can be combined into a single equation:

$$\Delta H = \Delta G + T\Delta S \tag{1.1}$$

where ΔH is the change in bond energy caused by the reaction, ΔG is the change in free energy, T is the absolute temperature in degrees Kelvin (i.e., temperature in Celsius plus $273.15°$), and ΔS is the change in entropy.

ΔG is also known as the *Gibb's free energy* after J. Willard Gibbs who introduced this concept in 1878. Consider a *reversible reaction* of the form:

$$2H_2 + O_2 \overset{K_{eq}}{\leftrightarrow} 2H_2O$$

where K_{eq} is known as the *equilibrium constant*. Assuming that the rate constant of the forward reaction is k and that of the reverse reaction is k_{-1}, then $K_{eq} = k/k_{-1}$. The Gibb's free energy for the forward reaction can be expressed as:

$$\Delta G = \Delta G^\circ + RT \ln\{([H_2O]^2)/([H_2]^2[O_2])\} \tag{1.2}$$

where ΔG° is known as the *change in standard free energy*, $R = 1.987$ cal/mol (calories per mole) is the *gas constant*, and T is the temperature in degrees Kelvin. The value of ΔG° is related to K_{eq} by the following equation:

$$\Delta G^\circ = -RT \ln K_{eq} \tag{1.3}$$

Combining Equations 1.2 and 1.3 results in:

$$\Delta G = RT \ln \frac{k_{-1}[H_2O]^2}{k[H_2]^2[O_2]} \tag{1.4}$$

In other words, ΔG is RT times the natural logarithm of the ratio of the rate of the reverse reaction over the forward reaction. When this value is negative, the forward reaction can occur spontaneously. When this value is positive, the reverse reaction can occur spontaneously. When this value is zero, the reaction is in a *steady state* (i.e., the forward and reverse reactions have equal rates of reaction).

From the above discussion, it would appear that chemical reactions with a positive free energy cannot occur. However, free energies of chemical reactions are additive. In other words, a reaction with a positive free energy can occur if there exists another reaction or reactions in the system that when their free energies are added together result in a cumulative negative free energy. One of the most common reactions that is used for this purpose is the *hydrolysis* of *adenosine triphosphate* (ATP) given by the equation below:

$$ATP + H_2O \leftrightarrow HPO_4^{2-} + ADP.$$

The forward reaction releases energy (i.e., has a negative free energy) while the reverse reaction stores energy. Coupling the forward reaction above with other reactions that have a positive free energy can allow these reactions to occur. These types of ATP reactions occur in all living organisms. Therefore, ATP is known as the *universal energy currency* of living organisms.

Even reactions with a negative free energy may not occur spontaneously as there is typically an energy barrier known as the *activation energy* that must be overcome first. For example, a collision between two molecules is typically necessary to get them close enough to break existing chemical bonds. The amount of these collisions can be increased by applying heat. Another mechanism is the use of an *enzyme*. An enzyme, or catalyst, is a reactant that accelerates a reaction without being consumed by the reaction. This text also uses the term *modifier* for such a chemical species to differentiate it from reactants which are consumed by reactions. It is typically the case that the amount of enzyme is much smaller than that of the other reactants, also known as *substrates*. Note that enzymes do not effect the overall free energy of the reaction, but rather they simply help the reaction overcome its activation energy barrier.

1.2 Macromolecules

While nearly 70 percent of a living organism is made up of water, the remainder is largely macromolecules often composed of thousands of atoms. There are four kinds of macromolecules: *carbohydrates, lipids, nucleic acids,* and *proteins.*

Carbohydrates are made up of carbon and water, so their chemical formula is basically $C_n(H_2O)_m$, where m and n are often equal or nearly so. Carbohydrates are also called sugars. An example carbohydrate is *glucose*. Carbohydrates are an important source of chemical energy that is used to power biological processes. The energy stored in their bonds is used to power everything from cell movement to cell division to protein synthesis. They also form part of the backbone for DNA and RNA.

Lipids are made up of mostly carbon and hydrogen atoms. They often have both a *hydrophilic* (water-loving) part and a *hydrophobic* (water-fearing) part. Their primary use is to form *membranes*. Membranes separate cells from one another and create compartments within cells as well as having other functions. Lipids make good membranes because their hydrophobic parts attract to form *lipid bi-layers* where the exterior allows water, but their interior repels water. This property allows the lipid bi-layers to form between areas containing water, but they do not allow water to easily pass through. Examples of simple lipids include fats, oils, and waxes.

Nucleic acids are the macromolecules that store information within living organisms. A nucleic acid, or *nucleotide*, is composed of a *chemical base* bound to a sugar molecule and a *phosphate* molecule. A sequence of nucleic acids connected by their phosphate molecules encodes the instructions to construct the product produced by a *gene*. Nucleic acids can be composed together in one of two forms: DNA (*deoxyribonucleic acid*) or RNA (*ribonucleic acid*). In DNA, the sugar molecule to which the base is bound is *deoxyribose*, while for RNA, it is *ribose*. Although RNA is used by a few viruses, most organisms use DNA to encode their genetic information. A strand of DNA is composed of four types of bases: *adenine* (A), *guanine* (G), *cytosine* (C), and *thymine* (T). Adenine and guanine are known as *purines* while cytosine and thymine are known as *pyrimidines*. These bases are each composed of oxygen, carbon, nitrogen, and hydrogen as shown in Figure 1.1. A strand of DNA is always read in one direction. One end of the DNA is called its *5' end*, which terminates in a 5' phosphate group (-PO4). The other end is called the *3' end*, which terminates in a 3' hydroxyl group (-OH). The genetic information is always read from the 5' to the 3' end. DNA is a *double-stranded* molecule as it consists of two strands running in opposite directions. The base pairs *A-T* and *G-C* are *complementary* in that if an A is found on one strand, then a T is found on the other strand in the same location. This feature is due to the fact that A bound to T and G bound to C are so strongly thermodynamically favored that any other combination is highly unlikely. This base pairing property allows one strand of DNA to serve as a *template* for synthesizing another complementary strand which greatly facilitates making copies of DNA sequences. The chemical makeup of this base pairing creates a force that twists the DNA into its well-known coiled double helix structure. RNA is very similar to DNA in that it is a chain of nucleotides with a 5' to 3' direction. However, it is typically found as a single-stranded molecule, it uses a ribose sugar molecule with a 2' hydroxyl group (-OH), and *uracil* (U) takes the place of the thymine nucleotide. RNA has many uses. For example, a strand of *messenger RNA* (mRNA) is created from genes that code for proteins during *transcription*. This mRNA molecule is used to carry the genetic information encoded in the DNA to the *ribosome*, the protein assembly machinery. The mRNA molecule is then used by the ribosome as a template to synthesize the protein that is encoded by the gene. There are several other types of RNA including *ribosomal RNAs* (rRNAs), *transfer RNAs* (tRNAs), and *small nuclear RNAs* (snRNAs).

Proteins are the basic building blocks of nearly all the molecular machinery of an organism. A cell may contain thousands of different proteins. Protein molecules are made from long chains of *amino acids*. There are 20 different kinds of amino acids found in living organisms. Genes in the DNA specify the order of the amino acids that make up the protein. This order in turn determines the protein's shape and function. During transcription, the code for a protein is transferred by using the DNA as a template to construct a strand of mRNA. This mRNA then proceeds to a ribosome where it is *translated* into a protein. A ribosome constructs a protein one amino acid at a time in the order specified by the *codons* in the mRNA. A codon is a group of three bases which specifies a particular amino acid using the *genetic code* shown in Table 1.1. The first codon (UUU) was associated with the amino acid phenylalanine by Nirenberg and Matthaei in 1961. Notice in the table that most amino acids are associated with more than one codon. These redundant codons often differ in the third position as is the case for the codons for Valine (i.e., GUU, GUC, GUA, and GUG). This redundancy provides robustness

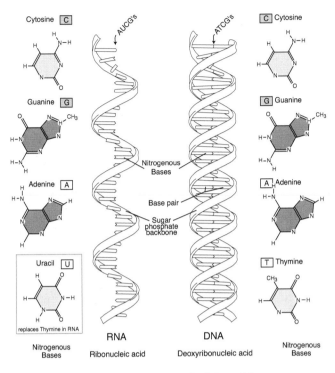

FIGURE 1.1: Nucleic acids
(courtesy: National Human Genome Research Institute).

in the face of *mutations* (i.e., small random changes) that occur naturally when DNA is replicated during cell division. The codons UAA, UAG, and UGA do not code for any amino acid. Instead, they are used to tell the ribosome when the protein is complete and translation should stop. After a protein is constructed, it folds into a specific three-dimensional configuration. The shape and position of the amino acids in this folded state determines the function of the protein. Therefore, understanding and predicting *protein folding* has become an important area of research. The structure of a protein is described in four levels as shown in Figure 1.2. The *primary structure* is simply the sequence of amino acids that make up the protein. The *secondary structure* is the patterns formed by the amino acids that are near to each other. Examples include *α-helices* and *β-pleated sheets*. The *ternary structure* is the arrangement of amino acids that are far apart. Finally, the *quaternary structure* describes the arrangement of proteins that are composed of multiple amino acid chains.

TABLE 1.1: The genetic code

	U	C	A	G
U	UUU Phenylalanine	UCU Serine	UAU Tyrosine	UGU Cysteine
	UUC Phenylalanine	UCC Serine	UAC Tyrosine	UGC Cysteine
	UUA Leucine	UCA Serine	UAA Stop	UGA Stop
	UUG Leucine	UCG Serine	UAG Stop	UGG Tryptophan
C	CUU Leucine	CCU Proline	CAU Histidine	CGU Arginine
	CUC Leucine	CCC Proline	CAC Histidine	CGC Arginine
	CUA Leucine	CCA Proline	CAA Glutamine	CGA Arginine
	CUG Leucine	CCG Proline	CAG Glutamine	CGG Arginine
A	AUU Isoleucine	ACU Threonine	AAU Asparagine	AGU Serineine
	AUC Isoleucine	ACC Threonine	AAC Asparagine	AGC Serineine
	AUA Isoleucine	ACA Threonine	AAA Lysine	AGA Arginine
	AUG Methionine	ACG Threonine	AAG Lysine	AGG Arginine
G	GUU Valine	GCU Alanine	GAU Aspartate	GGU Glycine
	GUC Valine	GCC Alanine	GAC Aspartate	GGC Glycine
	GUA Valine	GCA Alanine	GAA Glutamate	GGA Glycine
	GUG Valine	GCG Alanine	GAG Glutamate	GGG Glycine

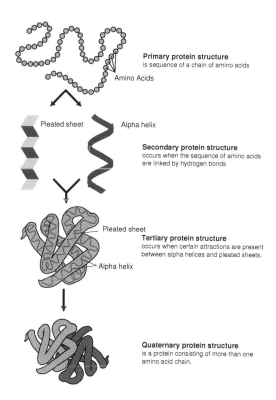

FIGURE 1.2: Protein structure
(courtesy: National Human Genome Research Institute).

1.3   Genomes

Although there are over 30 million types of organisms ranging in complexity from bacteria to humans, they all use the same basic materials and mechanisms to produce the building blocks necessary for life. Namely, information encoded in the DNA within its *genome* is used to produce RNA which produces the proteins that all organisms need. This amazing fact gives substantial support to the notion that all life began from a common origin. A genome is divided into genes where each gene encodes the information necessary for constructing a protein (or possibly an RNA molecule) using the genetic code shown in Table 1.1. Some of these proteins also control the timing for when other genes should produce their proteins as described later in Section 1.5.

The term gene, which was introduced in 1909 by the Danish botanist Wilhelm Johanssen, means the hereditary unit found on a *chromosome* where a chromosome is a linear DNA molecule. Genes, however, were discovered 50 years earlier by Gregor Mendel, though he called them *factors*. He was an Austrian monk who experimented with his pea plants in the monastery gardens. Since pea plants have both male and female organs, they normally self-fertilize. However, with the use of a pair of clippers, he was able to control the plants parents and ultimately their traits. It was a remarkable discovery that went largely ignored for nearly 50 years until the late 19th century when three researchers essentially duplicated his results. After studying the literature, they discovered Mendel's work and made it known to the public.

Initially, it was not completely accepted that genes are part of DNA. However, in 1953, James Watson and Francis Crick, with support from X-ray data from Rosalind Franklin and Maurice Wilkens, discovered the double helix structure of DNA (Watson and Crick, 1953). This discovery showed that DNA is composed of two strands composed of complementary bases. This base pairing idea shed light on how DNA could encode genetic information and be readily duplicated during cell division. Between 1953 and 1965, work by Crick and others showed how DNA codes for amino acids and thus proteins.

At the turn of this century, another major milestone was accomplished. In June 2000 at a White House press conference, the completion of a "working draft" of the human genome was announced. In February 2001, two largely independent drafts of the human genome were published (International Human Genome Sequencing Consortium, 2001; Venter *et al.*, 2001). Both drafts predicted that the human genome contains 30,000 to 40,000 genes (though now believed to be about 20,000 to 25,000). They made this estimate by using computer programs to analyze their sequence database to count *open reading frames*, (i.e., sequences of 100 bases without a stop codon), *start codons* such as ATG, specific sequences found at *splice junctions*, and *regulatory sequences*. Estimates have been improved in recent years using partial mRNA sequences to precisely locate genes by aligning the start and end portions with sequences in a DNA sequence database.

Although quite important, *structural genes* (DNA sequences that code for proteins) only make up about 1 percent of the 3 billion bases in the DNA for the human genome. The DNA also contains regulatory sequences that mark the start and end of genes and those used to switch genes on and off, but they also account for only a very small amount of the DNA. In fact, over 98 percent of our genome is called *"junk" DNA*, since it has no known function. About 40 to 45 percent of our genome is composed of *repetitive DNA*, short sequences often repeated 100s of times. Although they have no known role in the synthesis of proteins, they are an excellent *marker* for identifying people. Junk DNA may have once contained real genes that are no longer functional due to mutations, and may form a record of our evolutionary history.

In the human genome as well as that of all multicellular organisms, genes are not even continuous sequences. The *exons*, or coding portions of a gene, are broken up as shown in Figure 1.3 by *introns*, sequences that do not code for a protein. Although both exons and introns are transcribed into mRNA, the introns are snipped out before the mRNA is translated into a protein. The purpose of introns remains unclear, but they may serve as sites for *recombination* (i.e., large-scale rearrangements of the DNA molecule). To complicate matters further, about 40 percent of genes are *alternatively spliced*, meaning that the part that is considered introns may vary. The result is that a single gene can actually code for multiple proteins. One possible reason for alternative splicing is that it may reduce the chances of harmful mutations being passed to the next generation. If there are less genes, then there is a lower probability that a mutation affects an actual gene. If it does affect a gene, it likely affects the correct production of several proteins derived from alternative splices, thereby reducing the likelihood that the individual survives to produce a new generation. Extensive use of alternative splicing may also explain how humans with 20,000 to 25,000 genes are so much more complicated than the roundworm and fruit fly that have comparable genome sizes.

1.4   Cells and Their Structure

All living organisms are composed of *cells*. On the one hand, some organisms are composed of only a single cell. *Bacteria* are one such type of *unicellular* organism. On the other hand, *multicellular* organisms are composed of many cells. In fact, humans are estimated to be made up of more than 100,000,000,000,000 cells! Each cell must be capable of taking in nutrients, converting these into energy, eliminating waste products, growing and developing, responding to various stimuli from their environment, and forming a complete copy of itself (i.e., able to *reproduce*). Within each cell, the instructions for performing these tasks are contained in its genome.

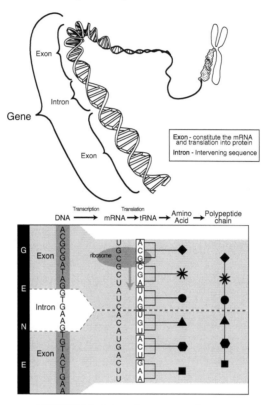

FIGURE 1.3: Introns and exons
(courtesy: National Human Genome Research Institute).

Life on earth began about 3.5 billion years ago with *prokaryotes*. Prokaryotes are unicellular organisms that lack a nuclear membrane and have few, if any, *organelles*. Most of the functions performed by organelles in higher organisms are performed by the *prokaryotic cell (plasma) membrane*. The major features of prokaryotic cells are shown in Figure 1.4(a). These features include external appendages such as the whip-like *flagella* for locomotion and the hair-like *pili (fimbriae)* for adhesion. The cell envelope consists of a *capsule*, a *cell wall*, and a *cell (plasma) membrane*. Finally, there is the *cytoplasm* that contains the cell's chromosome, ribosomes, and various sorts of *inclusions* (reserve deposits possibly membrane-bound of a variety of different substances). The DNA of the prokaryotic chromosome is organized in a simple circular structure. Prokaryotes do not develop or differentiate into multicellular organisms. Bacteria are the best known and most studied form of prokaryote. While some bacteria do grow in masses, each cell exists independently. Bacteria can inhabit almost any place on earth.

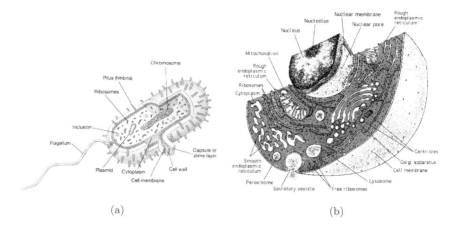

(a)   (b)

FIGURE 1.4: Major features of (a) a prokaryote and (b) a eukaryote (courtesy: National Human Genome Research Institute).

Eukaryotes appear in the fossil record about 1.5 billion years ago. Eukaryotes include fungi, mammals, birds, fish, invertebrates, plants, and complex single-celled organisms. Eukaryotic cells are also about 10 times larger than prokaryotes, and hence their volume is up to 1000 times larger. While eukaryotes use the same genetic code and metabolic processes as prokaryotes, they have a higher level of organizational complexity which allows for the development of multicellular organisms. Eukaryotes also include many membrane-bounded compartments, known as organelles, where metabolic activities occur.

The major features of eukaryotic cells are shown in Figure 1.4(b). First, there is the cell (plasma) membrane, which provides a barrier between a eukaryotic cell and its environment. This membrane is a complex composed of lipids, proteins, and carbohydrates. This membrane also contains a variety of molecules embedded within it which act as channels and pumps to transport various molecules into and out of the cell as well as to provide signals about the cell's environment. Next, there is the *cytoskeleton* which is composed of *microfilaments* (long thin fibers) and *microtubules* (hollow cylinders). The cytoskeleton gives a cell its shape and holds organelles in place. It also helps during *endocytosis* (i.e., the uptake of external materials) and can move parts of the cell to facilitate growth and movement. Lastly, there is the cytoplasm which is the large fluid-filled space inside the cell. While in prokaryotes, the cytoplasm does not have many compartments, in eukaryotes, it includes many organelles. The cytoplasm is essential to a cell as it serves many functions including dissolving nutrients, breaking down waste products, and moving materials around the cell.

The largest organelle in a eukaryote is the *nucleus* which contains the cell's genome organized as chromosomes. For example, the nucleus of a human cell has 23 chromosomes. In eukaryotes, the nucleus is the location where DNA replication and transcription takes place where as in prokaryotes, these functions occur in the cytoplasm. The nucleus has a spheroid shape and is enclosed by a double membrane called the *nuclear membrane*, which isolates and protects the genome from damage or interference from other molecules. During transcription (described in Section 1.5), mRNA is created to carry copies of part of the genome out of the nucleus to the cytoplasm where it is translated by a ribosome into the protein molecule specified by its sequence.

Another important organelle is the ribosome, which is a protein and RNA complex used by both prokaryotes and eukaryotes to produce proteins from mRNA sequences. A ribosome is a large complex composed of structural RNA and about 80 different proteins. During translation (also described in Section 1.5), ribosomes construct proteins by using the genetic code to translate an mRNA sequence into an amino acid sequence. Due to the importance of protein synthesis, there are 100s or even 1000s of ribosomes in a single cell.

Ribosomes are found either floating freely in the cytoplasm or bound to the *endoplasmic reticulum* (ER). The ER is a network of interconnected membranes that form channels within a cell. It is used to transport molecules that are either targeted for modifications or for specific destinations. There is a rough ER and a smooth ER. The rough ER is given its rough appearance by the ribosomes adhered to its surface. The mRNA for proteins that either stay in the ER or are exported from the cell are translated at these ribosomes. The smooth ER receives the proteins synthesized at the rough ER. Proteins to be exported from the cell are passed to the *Golgi apparatus* to be processed further, packaged, and transported to other locations.

Mitochondria and *chloroplasts* are organelles that generate energy. Mitochondria are self-replicating and appear in various shapes, sizes, and quantities. They have both an outer membrane that surrounds the organelle and an inner membrane with inward folds known as *cristae* that increase its surface area. Chloroplasts are similar, but they are only found in plants. They also have a double membrane and are involved in energy metabolism. In particular, they use *photosynthesis* to convert sun's light energy into ATP. Mitochondria and chloroplasts have their own genome which is a circular DNA molecule. The *mitochondrial genome* is inherited from only the mother. It is believed that this DNA may have come from bacteria that lived within the cells of other organisms in a *symbiotic* fashion until it evolved to become incorporated within the cell. Although the mitochondrial genome is small, its genes code for some important proteins. For example, the *mitochondrial theory of aging* suggests that mutations in mitochondria may drive the aging process.

Lysosomes and *peroxisomes* are responsible for degrading waste products and food within a cell. They are spherical, bound by a membrane, and contain *digestive enzymes*, proteins that speed up biochemical processes. Since these enzymes are so destructive, they must be contained in a membrane-bound compartment. Peroxisomes and lysosomes are similar, but peroxisomes can replicate themselves while lysosomes are made in the Golgi apparatus.

1.5 Genetic Circuits

Genes encoded in DNA are used as templates to synthesize mRNA through the process of transcription, and mRNA is converted to proteins via the process of translation as depicted in Figure 1.5. Humans have between 20,000 and 25,000 genes while mustard grass has 25,300 known genes. Therefore, it is clearly not just the number of genes that is indicative of an organism's complexity. Increased complexity can be achieved by the regulatory network that turns genes on and off. This network precisely controls the amount of production of a gene product, and it can also modify the product after it is made. Genes include not only *coding sequences* that specify the order of the amino acids in a protein, but also regulatory sequences that control the rate that a gene is transcribed. These regulatory sequences can bind to other proteins which in turn either *activate* (i.e., turn on) or *repress* (i.e., turn off) transcription. Transcription is also regulated through *post-transcriptional modifications*, *DNA folding*, and other feedback mechanisms. Transcriptional regulation allows mRNA to be produced only when the product is needed. This behavior is quite analogous to electrical circuits in which multiple input signals are processed to determine multiple output signals. Thus, in this text, these regulatory networks are known as genetic circuits.

DNA transcription is performed by the enzyme *RNA polymerase* (RNAP) as shown in Figure 1.6(a). Eukaryotic cells have three different RNA polymerases. Transcription is initiated when a subunit of RNAP recognizes and binds to the *promoter sequence* found at the beginning of a gene. A promoter sequence is a unidirectional sequence that is found on one strand of the DNA. There are two promoter sequences upstream of each gene, and the location and base sequence of each promoter site varies for prokaryotes and eukaryotes, but both are recognized by RNAP. A promoter sequence instructs RNAP both where to start synthesis of the mRNA transcript and in which direction. After binding to the promoter sequence, RNAP unwinds the double helix at that point and begins to synthesize a strand of mRNA in a unidirectional manner. This strand is complementary to one of the strands of the DNA, and so it is known as the *antisense* or *template strand*. The other strand is known as the *sense* or *coding strand*. The transcription process terminates when the RNAP reaches a stop signal. Transcription in prokaryotes is stopped by having a short region of G's and C's that fold in on themselves causing the RNAP to release the *nascent*, or newly formed, strand of mRNA. Termination of transcription in eukaryotes is not fully understood.

The ability of RNAP to bind to the promoter site can be either enhanced or precluded by other proteins known as *transcription factors*. These proteins recognize portions of the DNA sequence near the promoter region known as *operator sites*. Transcription factors can bind to these operator sites to either help RNAP bind to the promoter in the case of an *activator* or block RNAP

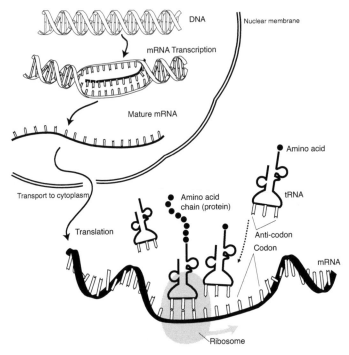

FIGURE 1.5: An overview of transcription and translation (courtesy: National Human Genome Research Institute).

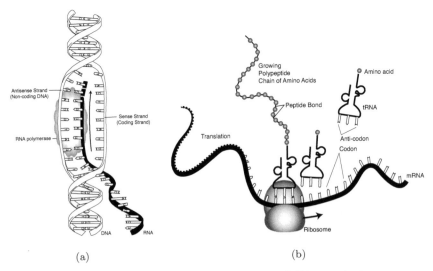

(a)   (b)

FIGURE 1.6: (a) Transcription and (b) translation (courtesy: National Human Genome Research Institute).

from binding to the promoter in the case of a *repressor*. In other words, an activator turns on (or enhances) gene expression while a repressor turns off (or reduces) gene expression. The effects of transcription factors can affect both adjacent genes known as *cis-acting* or distant genes known as *trans-acting*. Transcription can also be regulated by variations in the DNA structure and by chemical changes in the bases where the transcription factors bind. For example, *methylation* is a chemical modification of the DNA in which a *methyl group* (-CH3) is added. Methylation often occurs near promoter sites where there is cytosine preceded by guanine bases. Methylation affects the activity of a gene's expression in that it attracts a protein that binds to methylated DNA preventing the RNAP from binding to the promoter site.

The next step is protein synthesis from the mRNA by the translation process shown in Figure 1.6(b). Translation is performed by the ribosomes using tRNA. Each tRNA has an *anti-codon site* that binds to a particular sequence of three nucleotides known as a *codon*. A tRNA also has an *acceptor site* that binds to the specific amino acid for the codon that is associated with the tRNA. Ribosomes are made up of a *large subunit* and a *small subunit*. Translation is initiated when a strand of mRNA meets the small subunit. The large subunit has two sites to which tRNAs can bind. The *A site* binds to a new tRNA which comes bound to an amino acid. The tRNA then moves from the A site to the *P site* where it binds the tRNA to the growing chain of amino acids.

The translation process from mRNA into a polypeptide chain involves three steps: *initiation, elongation,* and *termination*. For prokaryotes, initiation begins when the ribosome recognizes and attaches at the *Shine-Delgarno sequence* (AGGAGGU) upstream from the the start codon (AUG) that also codes for methionine. Initiation for eukaryotes also usually begins at AUG, but sometimes with GUG. Next, a tRNA bound to methionine binds to this start signal beginning the elongation process. The ribosome shifts the first tRNA charged with methionine from the A site to the P site. A new tRNA that matches the next codons on the mRNA binds to the A site, and its amino acid is bound to the first. At this point, the first tRNA is released, the ribosome shifts the second tRNA now bound to a chain of two amino acids into the P site, and a new tRNA is drawn to the A site. This process continues until translation comes to one of the three *stop codons*, that signals that translation should move into the termination step. There is no tRNA that binds to the stop codon which signals the ribosome to split into its two subunits and release the newly formed protein and the mRNA template. At this point, the protein may undergo post-translational modifications while the mRNA is free to be translated again. A single mRNA transcript may code for many copies of a protein before it is degraded. It may even be transcribed by multiple ribosomes at the same time. Translation can be regulated by the binding of repressor proteins to the mRNA molecule. *Translational regulation* is heavily utilized during embryonic development and cell differentiation.

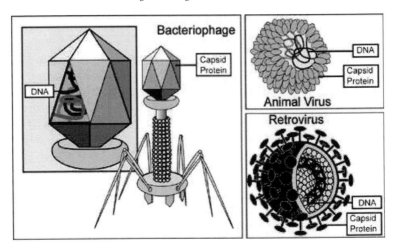

FIGURE 1.7:   Types of viruses
(courtesy: National Center for Biotechnology Information).

1.6   Viruses

Viruses cannot be classified as cells or to even be "living" since they do not have a metabolic system and they rely upon the host cells that they infect in order to reproduce. Viruses are simply genetic material, either DNA or RNA, surrounded by a protein package known as a *viral capsid*. Their genomes include genes to produce their protein package and those required for reproduction during infection. Since viruses must utilize the machinery and metabolism of a host cell to reproduce, they are known as *obligate intracellular parasites*. Before entering the host, the virus is known as a *virion*, or package of genetic material. A virion can enter a host through direct contact with an infected host or by a *vector*, or carrier. There are several types of viruses such as those shown in Figure 1.7. *Bacteriophages* are those that infect bacteria while *animal viruses* and *retroviruses* infect animals and humans.

The main goal of a virus is to replicate its genetic material. There are five main stages to virus replication: *attachment, penetration, replication, assembly,* and *release*. During attachment, the virus binds to receptors on the host's cell wall. During penetration, bacteriophages make a small hole in the cell wall and inject their genome into the cell leaving the virus capsid outside. Animal viruses and retroviruses, such as HIV, enter their host via endocytosis. During the replication stage, the virus begins the destructive process of taking over the cell and forcing it to produce new viruses. If the virus's genetic material is DNA, it can be replicated simply by the host when it copies its own DNA. If the virus uses RNA, it can be copied using *RNA replicase* in order to make a template to produce hundreds of duplicates of the original RNA. Retroviruses synthesize a complementary strand of DNA using the enzyme *reverse transcriptase*, which can then be replicated using the host cell machinery. During this stage, the virus also instructs the host to construct a variety of proteins which are necessary for the virus to reproduce. First, *early proteins* are produced which are enzymes needed for nucleic acid replication. Next, *late proteins* are produced that are used to construct the virus capsid. Finally, *lytic proteins* are produced, if necessary, to open the cell wall for exit. During the assembly stage, the virus parts are assembled into new viruses simply by chance or perhaps assisted by additional proteins known as *molecular chaperons*. During the release stage, assembled viruses leave the cell either by *exocytosis* or by *lysis*. For example, animal viruses instruct the host's endoplasmic reticulum to make *glycoproteins* that collect along the cell membrane forming exit sites from which the virus can leave the cell. Bacteriophages, on the other hand, typically must *lyse*, or break open, the cell to exit. These viruses have a gene that codes for the enzyme *lysozyme* that breaks down the cell wall causing it to swell and burst killing the host cell. The new viruses are then released into the environment.

1.7   Phage λ: A Simple Genetic Circuit

In 1953, Lwoff et al. discovered that a strain of *E. coli* when exposed to UV light lyse spewing forth viruses. He also found that some of the newly infected *E. coli* would soon lyse while others grow and divide normally until exposed to UV light. The virus is the phage λ shown in Figure 1.8. It is a bacteriophage that infects *E. coli*. The phage λ has two potential developmental strategies shown in Figure 1.9. The lysis strategy uses the machinery of the *E. coli* cell to copy its DNA and construct its protein coat. It then lyses the cell wall killing the cell and allowing the newly formed viruses to escape and infect other cells. The *lysogeny* strategy is a bit more subtle. In lysogeny, the virus inserts its DNA into the host cell DNA. Its DNA is then replicated through the normal process of cell division. If after perhaps many generations it detects the eminent demise of its host, it can revert to the lysis strategy to produce new viruses that escape to infect other cells. Since bacteria and their phages replicate quickly, phage λ has served as an excellent testbed for learning in the laboratory about how genes are turned on and off. Over the years, much has been learned about this simple genetic circuit. This section describes the decision circuit used by the phage λ to decide between these two potential developmental strategies. It serves to both illustrate the concepts involved in such genetic circuits as well as a running example to explain the analysis methods presented in later chapters.

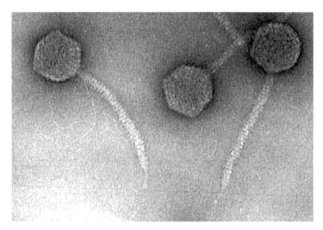

FIGURE 1.8: Bacteriophage Lambda (Phage λ), Electron Micrograph (negative stain) (courtesy of Maria Schnos and Ross Inman, Institute for Molecular Virology, University of Wisconsin, Madison).

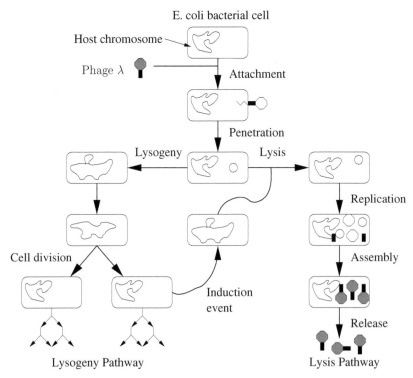

FIGURE 1.9: Phage λ developmental pathways.

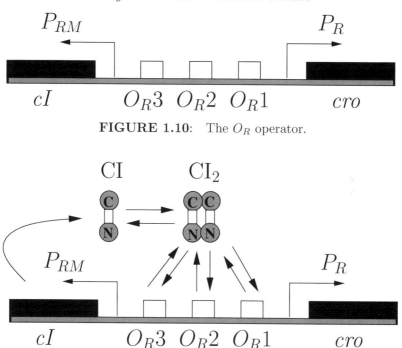

FIGURE 1.10: The O_R operator.

FIGURE 1.11: CI monomers are produced from the *cI* gene. Two CI monomers form a CI dimer which can bind to one of the O_R operator sites.

1.7.1 A Genetic Switch

The key element that controls the phage λ decision is the O_R (right) operator shown in Figure 1.10. The O_R operator is composed of three operator sites, O_R1, O_R2, and O_R3, to which transcription factors can bind. These operator sites are overlapped by two promoters. The P_{RM} promoter overlaps the operator site O_R3 and part of O_R2. RNAP can bind to P_{RM} initiating transcription to the left to produce mRNA transcripts from the *cI* gene. The P_R promoter overlaps O_R1 and the other part of O_R2. In this case, RNAP bound to this promoter transcribes to the right producing transcripts from the *cro* gene. These two promoters form a genetic switch since transcripts can typically only be produced in one direction at a time.

The *cI* gene codes for the CI protein, also known as the λ *repressor*. A single molecule, or *monomer*, of CI is composed of a *carboxyl* (C) and *amino* (N) domain connected by a chain of 40 amino acids. Two CI monomers react to form a dimer, CI_2. It is in dimer form that CI is attracted to the O_R operator sites in the phage's DNA. This process is shown in Figure 1.11. Similarly, the *cro* gene codes for the Cro protein which also dimerizes in order to bind to O_R operator sites as shown in Figure 1.12.

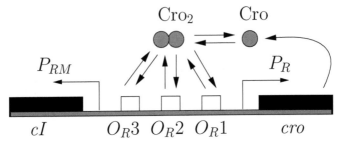

FIGURE 1.12: Cro monomers are produced from the *cro* gene. Two Cro monomers form a Cro dimer which can bind to one of the O_R operator sites.

Both CI_2 and Cro_2 can bind to the O_R operator sites to influence the direction of transcription. The molecule CI is known as the λ repressor because when CI_2 binds to O_R1, it turns off, or represses, the P_R promoter from which Cro production is initiated (see Figure 1.13(a)). CI_2 also serves as an activator when it is bound to O_R2. RNAP has only a weak *affinity*, or attraction, to the P_{RM} promoter. However, when CI_2 is bound to O_R2, it can draw RNAP to the promoter effectively activating the P_{RM} promoter as shown in Figure 1.13(b). Although shown in black-and-white terms, things in biology are rarely so clear-cut. When O_R3 and O_R2 are empty, nothing prevents RNAP to bind to P_{RM} and initiate transcription of CI, but RNAP binds here less often meaning that transcription proceeds at a reduced rate, known as the *basal rate*. The rate of transcription in the scenario shown in Figure 1.13(b) is known as the *activated rate*. The final possibility is that CI_2 is bound to O_R3 which does block RNAP from binding to P_{RM}, so it is completely off in this case. It has no effect, however, on P_R. While the P_{RM} promoter producing transcripts for the CI molecule is initially inactive, the P_R promoter producing transcripts for the Cro molecule is initially active as shown in Figure 1.13(d). Its promoter has a stronger affinity for RNAP. After Cro is produced, Cro_2 can bind to O_R3 preventing even basal production from P_{RM} as shown in Figure 1.13(e). Finally, if Cro_2 happens to also bind to either operator sites O_R1 or O_R2, it represses its own production (see Figure 1.13(f)).

While CI_2 and Cro_2 can bind to any of the three operator sites at any time, they have a different affinity to each site. The CI_2 has its strongest affinity to the O_R1 operator site, next to the O_R2 site, and finally to the O_R3 site. Therefore, as CI concentration increases, the likely configurations in which CI_2 is to be found bound to O_R are as shown in Figure 1.14. In other words, CI_2 first turns off P_R, then activates P_{RM}, and finally, represses its own production. The Cro_2 has the reverse affinity as shown in Figure 1.15. Namely, it is likely to first bind to O_R3 to turn off CI production. Next, it can either bind to O_R1 or O_R2 interfering with Cro production. Finally, in very high concentration, Cro_2 can be found bound to all three sites.

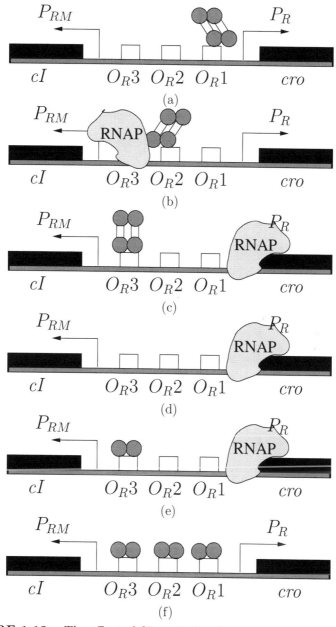

FIGURE 1.13: The effect of CI_2 and Cro_2 bound to each operator site. (a) CI_2 bound to O_R1 turns off P_R. (b) CI_2 bound to O_R2 turns on P_{RM}. (c) CI_2 bound to O_R3 turns P_{RM} completely off while P_R is on. (d) P_R is active when all operator sites are empty. (e) Cro_2 bound to O_R3 turns P_{RM} completely off. (f) Cro_2 also bound to O_R1 or O_R2 turns P_R off.

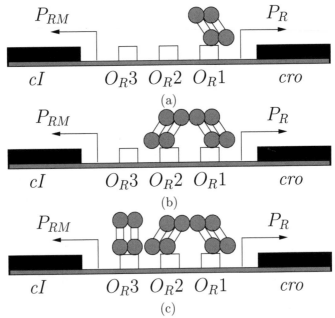

FIGURE 1.14:   Likely position of CI_2 bound to O_R versus CI_2 concentration. (a) At low concentration, CI_2 is most likely bound to O_R1. (b) Due to cooperativity, a second CI_2 molecule quickly binds to O_R2. (c) At high concentrations of CI_2, it likely also binds to O_R3.

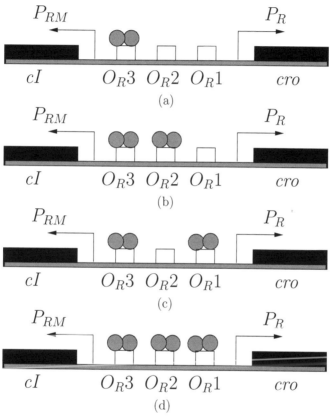

FIGURE 1.15: Likely position of Cro_2 bound to O_R versus Cro_2 concentration. (a) At low concentration, Cro_2 is most likely bound to O_R3. At moderate concentration, it is equally likely to be also bound to (b) O_R2 or (c) O_R1. (d) At high concentrations, it is likely bound to all three operator sites.

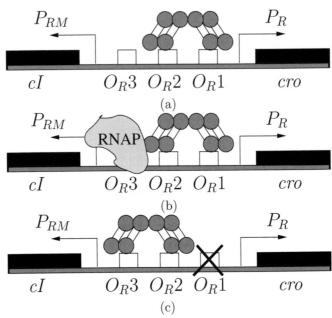

FIGURE 1.16: Cooperativity of CI_2 binding. (a) A CI_2 molecule bound
to O_R1 bends over to attract another CI_2 to O_R2. (b) A CI_2 molecule
bound to O_R2 attracts RNAP to P_{RM}. (c) CI_2 molecules can bind
cooperatively at O_R3 and O_R2 when O_R1 is disabled by a mutation.

   While it is possible to find a single CI_2 molecule bound to either O_R2 or
O_R3, it is highly unlikely. As mentioned above, the first dimer to bind to O_R
typically binds to O_R1. Next, this dimer helps attract another CI_2 molecule
onto the O_R2 site. It does this by bending over such that one carboxyl domain
from each dimer touch as shown in Figure 1.16(a). This effect is known as
cooperativity, and it is so strong that it appears almost as if the two dimers
bind simultaneously. The two dimers bound in this way have a dual effect
as shown in Figure 1.16(b). Namely, they repress production of Cro and
they activate production of CI. While not often found in *wild-type* (i.e., non-
mutated), the cooperativity effect can be found between dimers bound to O_R2
and O_R3 when the O_R1 operator site is disabled through mutation as shown
in Figure 1.16(c).
   The effect of cooperativity is that the repression of Cro by CI becomes
more switch like. Figure 1.17 shows the percent of repression of the P_R promoter (i.e., Cro production) versus CI concentration for both the wild-type λ
switch and one without cooperativity. The one without cooperativity is also
controlled by CI monomers rather than dimers in that dimerization is also a
form of cooperativity. A lysogen has approximately 4×10^{-7} M of CI. The
cooperative switch is very stable to small perturbations around this value.

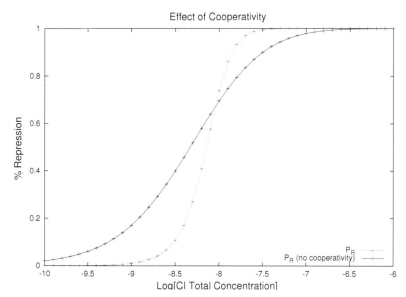

FIGURE 1.17: Effect of cooperativity of CI_2 molecules on repression of Cro production.

However, when the concentration of CI drops significantly, the repression of Cro production is rapidly released. In the non-cooperative switch, repression drops off much more gradually, making the transition to lysis much less sharp.

Given the above discussion, at low to moderate concentrations of CI and Cro, there are three common configurations. First, there may be no molecules bound to O_R. In this case, Cro is produced at its full rate of production while CI is only produced at its low basal rate. Second, CI_2 molecules may be found bound to O_R1 and O_R2. In this case, Cro production is repressed and CI production is activated. Third, a Cro_2 molecule may be bound to O_R3. In this case, CI cannot be produced even at its basal rate while Cro continues to be produced. Therefore, the feedback through the binding of the products as transcription factors coupled with the affinities described makes the O_R operator behave as a genetic bistable switch. In one state, Cro is produced locking out production of CI. In this state, the cell follows the lysis pathway since genes downstream of Cro produce the proteins necessary to construct new viruses and lyse the cell. In the other state, CI is produced locking out production of Cro. In this state, the cell follows the lysogeny pathway since proteins necessary to produce new viruses are not produced. Instead, proteins to insert the phage's DNA into the host cell are produced.

In the lysogeny state, the cell develops an immunity to further infection. The *cro* genes found on the DNA inserted by further infections of the λ virus are also shut off by CI_2 molecules that are produced by the first virus to

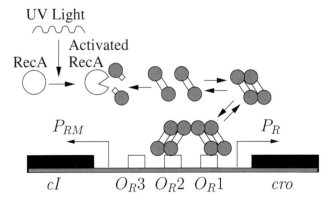

FIGURE 1.18:   The induction mechanism. UV light activates RecA
which can cleave CI monomers. The cleaved CI monomers are unable to
dimerize and bind to O_R1 which reduces the concentration of CI_2 molecules
allowing Cro production to begin.

commit to lysogeny. Once a cell commits to lysogeny, it becomes very stable
and does not easily change over to the lysis pathway. An *induction event*
is necessary to cause the transition from lysogeny to lysis. For example, as
described earlier, *lysogens* (i.e., cells with phage λ DNA integrated within
their own DNA) that are exposed to UV light end up following the lysis
pathway. UV light creates DNA damage that is potentially fatal to the cell.
In response, the cell activates the protein RecA which can help repair DNA
damage. It also has the effect of cleaving the CI monomer into two parts as
shown in Figure 1.18. This inactivates the CI molecule making it incapable
of forming a dimer and binding to O_R. As the concentration of complete CI
molecules diminish, the cell is unable to maintain the lysogeny state. Cro
production begins again moving the cell into the lysis pathway.

1.7.2   Recognition of Operators and Promoters

As described above, the proteins CI_2 and Cro_2 bind to operator sites which
are specific DNA sequences that are 17 base pairs long. How do these proteins locate these sequences from amongst the millions within the bacteria?
The second column of Table 1.2 shows the DNA sequences for the three O_R
operator sites as well as three more O_L (left) operator sites that also bind to
CI_2 and Cro_2. The top line of each row is one strand of the DNA while the
bottom line is the complementary strand. The base pair shown in lower case
represents the midpoint of the sequence. Observing from this midpoint, one
finds that a strand on one side of the midpoint is nearly symmetric with the
complementary strand on the other side. This may not, however, be readily
obvious. Therefore, in the third column, the top of each row is the first half

TABLE 1.2: Near symmetry in the operator sequences

Op	Operator sequences	Operator half sequences
O_R1	T A T C A C C G c C A G A G G T A	T A T C A C C G c
	A T A G T G G C g G T C T C C A T	T A C C T C T G
O_R2	T A A C A C C G t G C G T G T T G	T A A C A C C G t
	A T T G T G G C a C G C A C A A C	C A A C A C G C
O_R3	T A T C A C C G c A A G G G A T A	T A T C A C C G c
	A T A G T G G C g T T C C C T A T	T A T C C C T T
O_L1	T A T C A C C G c C A G T G G T A	T A T C A C C G c
	A T A G T G G C g G T C A C C A T	T A C C A C T G
O_L2	T A T C T C T G g C G G T G T T G	T A T C T C T G
	A T A G A G A C c G C C A C A A C	C A A C A C C G c
O_L3	T A T C A C C G c A G A T G G T T	T A T C A C C G c
	A T A G T G G C g T C T A C C A A	A A C C A T C T
Con.		T_9 A_{12} T_6 C_{12} A_9 C_{11} C_7 G_9 C_5
		C_2 C_3 T_2 T_1 T_4 T_2 T_1
		A_1 A_3 C_1 G_1 C_1

of the top strand, and the bottom is the second half of the complementary strand reversed. The last row accumulates frequencies of each base pair in each position. The most likely entries form the following *consensus sequence*:

TATCACCGcCGGTGATA

ATAGTGGCgGCCACTAT

These frequencies show that many entries are highly preserved. For example, there is always an A in the 2nd (and 16th) position, a C in the 4th (and 14th) position, and nearly always a C in the 6th (and 12th) position. There are, however, some differences. It is these differences that cause the differences in affinity for CI_2 and Cro_2 for the different operators. Notice that the first half of the operator sites O_R1 and O_R3 agree perfectly with the consensus sequence. The second half, however, has several differences which lead to different affinities for CI_2 and Cro_2.

To see how CI_2 and Cro_2 recognize these operator sites differently, it is necessary to consider their structures in more detail. The amino domains of a CI_2 molecule are composed of five α-helices that are responsible for recognizing where to bind onto to the DNA. One α-helix in particular plays a critical part in this role as it rests inside the *major groove* of the DNA (the larger gap in the DNA helix) when bound. There is a comparable α-helix in the Cro_2 molecule. The amino acids that make up these α-helices for CI_2 and Cro_2 are shown in Figure 1.19(a) and (c). These figures also show the attractions between these amino acids and the bases within the second half of the sequences for O_R1 and O_R3 (note that they are reversed and inverted from how they are shown in Table 1.2). Both CI_2 and Cro_2 begin with the amino acid glutamine (gln) which is attracted to the A-T base pair in the second position. They also agree in the second amino acid serine (ser) which is drawn to the C-G base pair in the fourth position. This commonality shows why CI_2 and Cro_2 are both attracted to these operator sites. They differ, however, in the remaining amino acids. For example, Cro_2 includes asparagine (asn) which is drawn

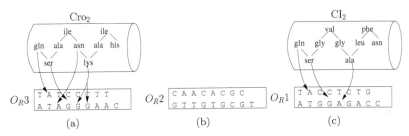

FIGURE 1.19:   Amino acid-base pair interactions. (a) CI_2 with O_R1 sequence. (b) O_R2 sequence. (c) Cro_2 with O_R3 sequence.

TABLE 1.3: λ promoters

	-35		-10
Consensus	T T G A C A	- 17bp -	T A T A A T
λP_{RM}	T A G A T A	- 17bp -	T A G A T T
λP_R	T T G A C T	- 17bp -	G A T A A T

to the T-A base pair in position three, and lysine (lys) that is drawn to the C-G base pairs in positions five and six. Similarly, CI_2 includes alanine (ala) which is attracted to the T-A base pair in the fifth position of O_R1. Note that the fifth position of O_R3 is C-G which makes for a less favorable bond for CI_2. These attractions make Cro_2 bond more favorably with O_R3 and less favorably with O_R1. The case is the reverse for O_R1. Note that as shown in Figure 1.19(b), O_R2 has a common base pair in the third position with O_R3 and a common base pair in the fifth position with O_R1. This makes O_R2 rest in the middle in terms of affinity for both CI_2 and Cro_2.

Another important characteristic of the genetic switch is that RNAP can readily bind to the P_R promoter to produce Cro, but the P_{RM} promoter for CI production must be activated by having CI_2 bound to O_R2. RNAP like transcription factors must locate a particular sequence associated with a promoter on which to bind. There are many promoters each associated with one or more genes on a strand of DNA. Considering the sequences associated with these promoters, one can generate a consensus sequence indicating the most likely (i.e., best) promoter configuration. Table 1.3 shows a portion of this consensus sequence. It turns out that the most important part of the sequence are the 6 base pairs located near the -35 and -10 positions where -35 means 35 base pairs away from the start of the gene. Comparing these portions of the consensus sequence with the same portions of the P_{RM} promoter and the P_R promoter, it is found that the P_R promoter is a better match to the consensus sequence than P_{RM} differing in only two bases while P_{RM} differs in four. For this reason, RNAP can easily bind to P_R and must typically be activated to bind to P_{RM}.

1.7.3   The Complete Circuit

Let us now consider how the genetic switch works with the rest of the λ circuit to decide between taking the lysis or lysogeny pathway. The first step is *circularization*. Phage λ's DNA when it enters the host is in the form of a linear strand. At each end of this strand is 12 bases of single stranded DNA known as *cohesive*, or sticky, *ends* which join to form a circular strand of DNA. This circular genome for λ is shown in Figure 1.20. The location marked *cos* is the location of the cohesive ends. Continuing in a clockwise direction, the next 22 genes encode proteins that construct the head and tail of the virus. The next group of ten genes includes two genes *int* and *xis* which are used to integrate the λ genome within the host's genome during lysogeny and excise it during induction. The site labeled *attp*, or *attachment site*, is where the DNA is split when it in integrated within the *E. coli* genome during lysogeny. Next, comes five individual genes labeled, *cIII, N, cI, cro,* and *cII* which are used to make the decision between lysis and lysogeny as described below. The next two genes, *O* and *P*, are used during replication of the λ genome. The *Q* gene is used during the lysis pathway. Finally, the three genes labeled lysis are used to lyse, or open, the bacteria.

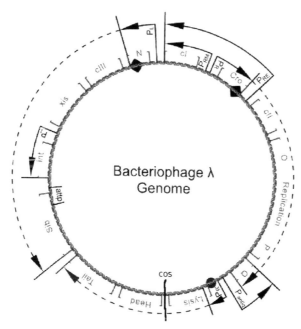

FIGURE 1.20:   The λ genome (courtesy: http://en.wikipedia.org/wiki/Lambda_phage).

The λ genome also includes seven promoters, which include the P_{RM} and P_R promoters, discussed earlier, that are responsible for starting transcription of the *cI* and *cro* genes, respectively. Transcripts from P_{RM} are always terminated immediately after transcribing the *cI* gene. Transcripts from P_R, however, encounter a *terminator switch* after transcribing the *cro* gene and can continue to transcribe *cII*, *O*, *P*, and *Q* genes. This potential transcription is indicated with a dashed line. The promoter P_L begins transcription for the *N* gene. Transcripts beginning at P_L also encounter a terminator switch and can potentially continue to transcribe *cIII*, *xis*, and *int* genes. The *int* gene can also be transcribed starting from the P_I promoter. The $P_{R'}$ promoter transcribes the genes for the proteins needed for the lysis pathway. The P_{antiq} promoter produces reverse transcripts for the gene *Q*. The mRNA for these transcripts can bind to the mRNA from forward transcripts for the gene *Q* effectively preventing synthesis of the protein *Q*. Finally, the P_{RE} promoter can produce transcripts of the *cI* gene as well as reverse transcripts for the *cro* gene. The remainder of this section describes this complete circuit in more detail. The genes in the λ genome can be divided into those which are active *very early*, *early*, or *late* as shown in Figure 1.21.

The *N* and *cro* genes are the only ones that are active very early. The other genes such as *cI* do produce transcripts, but only at a low basal rate initially. Recall that a buildup of Cro can trigger the lysis pathway to be taken. The protein N is known as an *anti-terminator*, and its role is illustrated in Figure 1.22. RNAP that binds to the promoter P_L (known as the left promoter since transcripts move from the left of the gene *cI*) transcribes the gene *N*, and it then hits a terminator switch as shown in Figure 1.22(a). This action blocks transcription about 80 percent of the time such that all genes downstream of the switch do not get transcribed, and the proteins that they code for do not get produced. However, after some of the protein N is produced, it is now possible that as the RNAP passes over the *nut site*, NUT_L, (shown as a diamond in the figures) the protein N may bind to the RNAP as shown in Figure 1.22(b). The result is that the RNAP now passes over the terminator switch and continues to transcribe the remaining genes as shown in Figure 1.22(c). As mentioned above, there is another terminator switch that blocks transcripts to the right, and it is also controlled by the N protein.

The transition from the very early to the early stage is marked by a buildup of the protein N. As just described, the protein N closes the terminator switches allowing transcripts from the P_R and P_L promoters to transcribe additional genes (see the dashed arrows in Figure 1.21). Transcripts beginning from the P_L promoter now proceed past its terminator switch to transcribe the *cIII*, *xis*, and *int* genes. These transcripts also continue on to transcribe an important non-genetic portion of the DNA known as *sib*. The *sib* portion of the mRNA forms a hairpin as shown in Figure 1.23(a). This hairpin attracts the enzyme *RNase3* which degrades the hairpin and allows for other enzymes to come along and gradually destroy the rest of the mRNA as shown

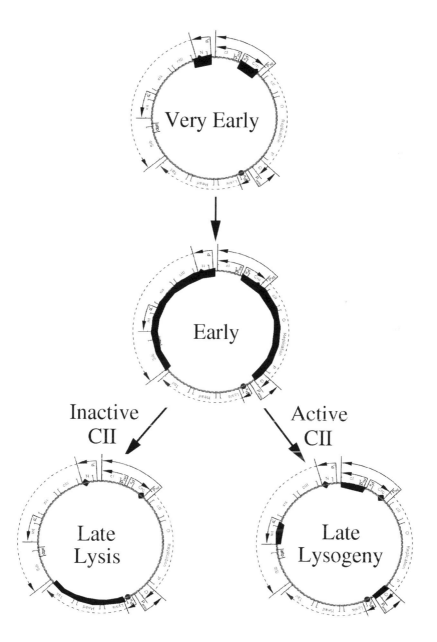

FIGURE 1.21: Patterns of gene expression.

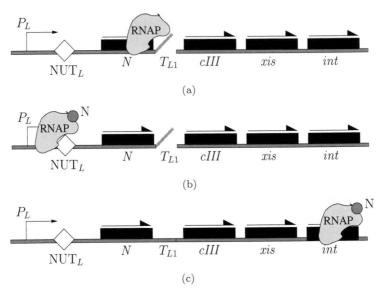

FIGURE 1.22: The action of N. (a) 80 percent of the transcripts stop at the terminator switch. (b) N can bind to RNAP at NUT_L. (c) RNAP with N bound to it always proceeds over the terminator switch and transcribes the *cIII*, *xis*, and *int* genes.

in Figure 1.23(b). Since the Int portion is destroyed before the Xis portion, more of the Xis protein is allowed to be synthesized than Int as shown in Figure 1.23(c). The Int protein is used to integrate the λ genome within the host's DNA while the Xis protein is used to extract it. An excess of Xis prevents the DNA from being integrated, so it prevents lysogeny. The buildup of N also allows for more transcripts from P_R to proceed past its terminator to transcribe the *cII*, *O*, *P*, and *Q* genes.

There are two potential sets of late genes depending on whether lysis or lysogeny has been chosen. The key to this decision is the activity of the protein CII. This protein activates the P_{RE} promoter which gives a jump-start to the production of the protein CI. Without production from P_{RE}, the CI protein has a difficult time to buildup. With it, positive feedback in P_{RM} further increases CI production and locks out further Cro production driving the virus down the lysogeny pathway. The activity of CII is determined by environmental factors. Bacterial *proteases* (enzymes that degrade proteins) attack and destroy CII readily, making it very unstable. Growth in a nutrient rich medium activates these proteases whereas starvation has the opposite effect. For this reason, well-fed cells tend towards lysis. The protein CIII produced from the P_L promoter protects CII from degradation promoting lysogeny. The production of CIII is limited by the terminator switch. Therefore, production of the anti-terminator, N, is needed to enhance CIII production. When

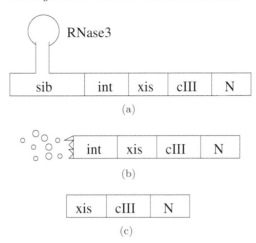

FIGURE 1.23: Retroregulation of Int. (a) The *sib* portion forms a hairpin that attracts the *RNase3* enzyme. (b) This degrades the mRNA starting at the hairpin. (c) The Int portion of the transcript is degraded first resulting in an excess of Xis.

more phages infect a single *E. coli* cell, then there are more *N*, *cII*, and *cIII* genes available for transcription leading to higher concentrations of the resulting proteins. Therefore, higher *multiplicities of infection* lead to a higher probability of lysogeny.

The late genes active for the lysis case are shown in Figure 1.21. The buildup of Cro shuts off the P_L and P_R promoters stopping production from all the early genes. At this point, the O and P proteins constructed during the early stage begin to facilitate the replication of the λ genome to be inserted into each new viral capsid. The protein Q that also built up during the early stage activates the $P_{R'}$ promoter. In particular, the Q protein can bind to RNAP at the *Qut site* (shown as a circle in Figure 1.21). This modification of the RNAP allows transcription from the $P_{R'}$ promoter to continue past its terminator switch. As a result, genes are transcribed that code for proteins to construct the heads and tails of new viral capsids as well as those necessary to lyse the cell.

The late genes for the lysogeny case are shown in Figure 1.21. The CII protein activates three additional promoters, P_{antiq}, P_{RE}, and P_I. The mRNA transcripts produced from the P_{antiq} promoter are complementary to the transcripts for the gene Q. Therefore, these transcripts can bind to the transcripts for the gene Q preventing them from being translated into Q proteins and helping to prevent the activation of $P_{R'}$ that is needed by the lysis pathway. The promoter P_{RE} produces transcripts of the gene *cI*. These initial CI molecules are then free to dimerize and bind to O_R to activate further production of CI starting from the promoter P_{RM} and repress production of Cro as

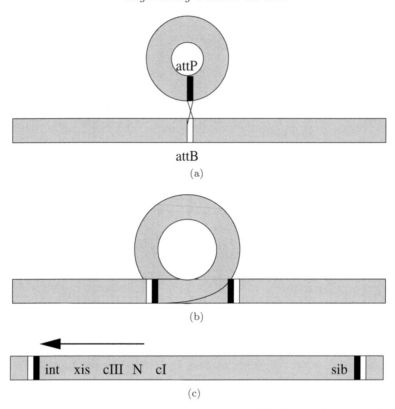

FIGURE 1.24: Integration and induction. (a) The *attP* site on the λ
genome lines up with the *attB* site on the host genome. (b) The λ genome
splits at this location and then rejoins with the host genome. (c) During
induction, the P_L promoter becomes active, but the *sib* region is not
transcribed as it is on the other side of the integrated genome.

described above. The promoter P_I is located in the middle of the *xis* gene, so
the transcripts that it produces do not produce the Xis protein. Production
from this promoter results in an excess of the Int protein that is needed to
insert the phage chromosome into the host's DNA.

The *integration* process begins with the phage attachment site (*attP*) coming into contact with the bacterium's attachment site (*attB*) as shown in
Figure 1.24(a). The λ genome splits at this location and then rejoins with
the host's genome as shown in Figure 1.24(b). Finally, this process can be
reversed during induction. Induction requires both the proteins Int and Xis.
During induction, transcripts are produced from the P_L promoter to produce
the proteins N, CIII, Xis, and Int. This time, however, it does not transcribe
the *sib* region as it is located on the other side of the *attP* site and ends up
on the other side of the integrated genome as shown in Figure 1.24(c).

1.7.4 Genetic Circuit Models

This subsection describes the chemical reaction network model for a portion of the phage λ decision circuit which is involved in CI production from P_{RE} and CII production from P_R. The chemical reaction network model for the complete phage λ decision circuit is given in an appendix at the end of this chapter. This portion of this genetic circuit is shown in Figure 1.25(a). Assuming initially that there are no CI or CII molecules, CII production is initiated from the P_{RE} promoter while CI production from P_R is only at a low basal rate. As CII builds up, it binds to the O_E operator site and activates production of CI molecules from P_{RE}. These CI molecules dimerize and bind to the O_R operator sites repressing further production of CII. Over time CII degrades, reducing the production rate of CI. Finally, after CI degrades, the system has returned to the initial state. This genetic circuit can be represented graphically as shown in Figure 1.25(b) in which it is shown that the CII protein activates CI production through the promoter P_{RE} while the CI protein represses CII production through the promoter P_R.

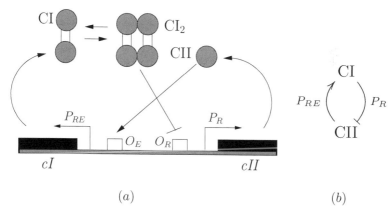

(a)   (b)

FIGURE 1.25: (a) Genetic circuit and (b) graphical model for a portion of the phage λ decision circuit which is involved in CI production from P_{RE} and CII production from P_R.

While there are a variety of different ways to model a genetic circuit using chemical reactions, the following procedure is systematic and produces a fairly reasonable model.

1. Create a species for RNAP as well as for each promoter and protein.

2. Create degradation reactions for each protein.

3. Create open complex formation reactions for each promoter.

4. Create dimerization reactions, if needed.

5. Create repression reactions for each repressor.

6. Create activation reactions for each activator.

This procedure is illustrated using the simple genetic circuit shown in Figure 1.25. For this example, step 1 adds species to the model for RNAP, promoters P_R and P_{RE}, and proteins CI and CII. The initial number of RNAP molecules has a default value of 30 while the initial number of promoters, P_R and P_{RE}, is 1. The initial number of molecules for the proteins, CI and CII, is 0. Step 2 creates degradation reactions for CI and CII as shown in Figure 1.26 in which CI and CII are reactants and there are no products. As shown in Figure 1.26, each species is given the same degradation rate, k_d, which has a default value of 0.0075 sec^{-1}, but a separate value may be specified for each species, if desired. Step 3 creates a reversible reaction (denoted by a double arrow in the diagram) for each promoter which models the binding of RNAP to the promoter to form a complex with default equilibrium constant of K_o. For each of the resulting complexes, another reaction is added with this complex as a modifier that models transcription and translation. The result of step 3 for this example is shown in Figure 1.27. Note that in this example, different constants, K_{o1} and K_{o2}, are used for the binding of RNAP to the promoters. Also, CI is only produced at a low basal rate, k_b, from P_{RE} because it needs to be activated while CII is produced at its full rate, k_o, from P_R. Finally, the value np indicates the average number of proteins produced per mRNA transcript. Step 4 creates a reversible reaction to form dimers for each species that must first dimerize to act as a transcription factor, as is the case for the CI molecule in this example (see Figure 1.28). Step 5 creates a reversible reaction that binds each repressor species to the promoter that it represses preventing it from being able to bind to RNAP. Note that the value nc specifies the number of molecules that must bind to cause repression. In this example, only one CI_2 molecule is needed to repress production of the CII molecule from P_R as shown in Figure 1.29. Finally, Step 6 adds a reversible reaction that binds each activator species with RNAP to the promoter that is activated. Note that the value na specifies the number of molecules that must bind to cause activation. The resulting complex is then used as a modifier in a reaction to produce a species at an activated rate of production. In this example, one CII molecule activates production of the CI molecule by binding to P_{RE} as shown in Figure 1.30. Putting this all together results in the complete reaction-based model for this example shown in Figure 1.31. This model has 10 species and 10 reactions of which 5 are reversible.

1.7.5 Why Study Phage λ?

Phage λ has been the subject of study for over 50 years now. It is one of, if not the best, understood genetic circuit. The phage λ is an excellent illus-

$$\mathrm{CI} \xrightarrow{k_d} ()$$
$$\mathrm{CII} \xrightarrow{k_d} ()$$

Constant	*Value*
k_d	$0.0075\ \mathrm{sec}^{-1}$

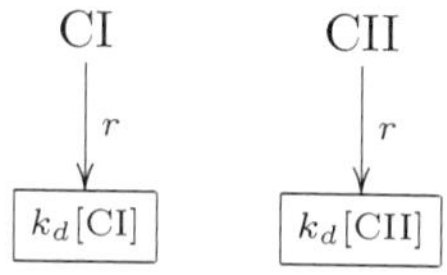

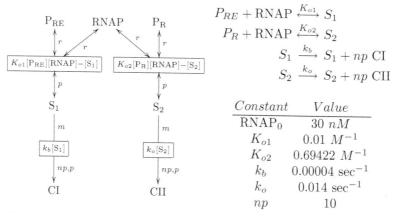

FIGURE 1.26: Degradation reactions for CI and CII.

$$P_{RE} + \mathrm{RNAP} \xleftrightarrow{K_{o1}} S_1$$
$$P_R + \mathrm{RNAP} \xleftrightarrow{K_{o2}} S_2$$
$$S_1 \xrightarrow{k_b} S_1 + np\ \mathrm{CI}$$
$$S_2 \xrightarrow{k_o} S_2 + np\ \mathrm{CII}$$

Constant	*Value*
RNAP_0	$30\ nM$
K_{o1}	$0.01\ M^{-1}$
K_{o2}	$0.69422\ M^{-1}$
k_b	$0.00004\ \mathrm{sec}^{-1}$
k_o	$0.014\ \mathrm{sec}^{-1}$
np	10

FIGURE 1.27: Open complex formation reactions for P_R and P_{RE}.

$$2\mathrm{CI} \xleftrightarrow{K_d} \mathrm{CI}_2$$

Constant	*Value*
K_d	$0.1 M^{-1}$

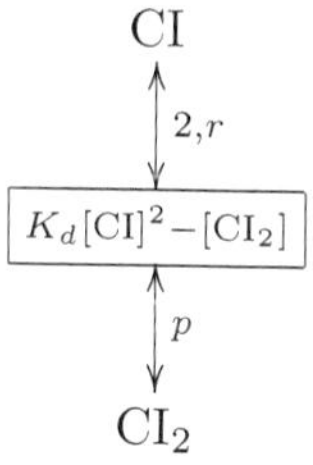

FIGURE 1.28: Dimerization reactions for CI.

$$P_R + nc\ \mathrm{CI}_2 \xleftrightarrow{K_r} S3$$

Constant	*Value*
K_r	$0.2165\ M^{-nc}$
nc	1

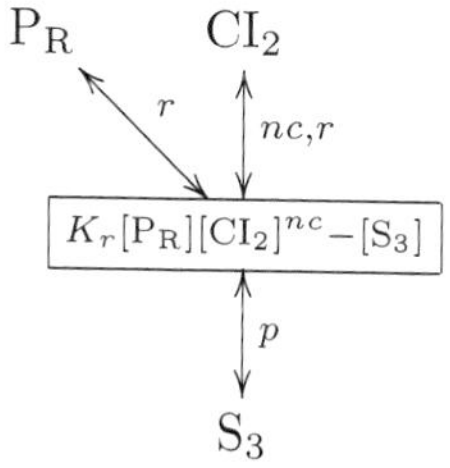

FIGURE 1.29: Repression reactions for P_R.

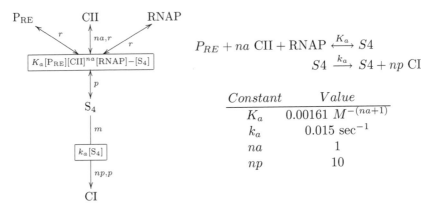

$$P_{RE} + na\ CII + RNAP \xrightleftharpoons{K_a} S4$$
$$S4 \xrightarrow{k_a} S4 + np\ CI$$

Constant	Value
K_a	$0.00161\ M^{-(na+1)}$
k_a	$0.015\ \text{sec}^{-1}$
na	1
np	10

FIGURE 1.30:   Activation reactions for P_{RE}.

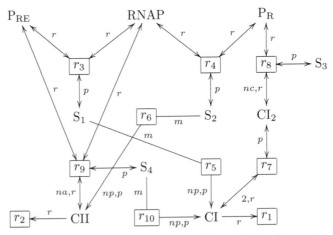

FIGURE 1.31:   Complete reaction-based model for the portion of the phage λ decision circuit which is involved in CI production from P_{RE} and CII production from P_R.

tration of a genetic circuit that is able to analyze its environmental situation and make an informed decision between two competing pathways. There are numerous systems to which parallels can be drawn. These range from other phages with similar mechanics to bacteria with circuits that must respond properly to various forms of stress to even circuits involved in development and cell differentiation. Genes from phage λ are even used in *synthetic biology* where DNA is produced to perform particular functions. For all these reasons, phage λ makes an excellent testbed for trying new ideas. Virtually every modeling method that has been developed for genetic circuits has been applied to phage λ. This text, therefore, also uses it as a running example.

1.8 Sources

Numerous sources were used during the development of this chapter which also make for excellent references for further reading. One such resource that inspired much of the material in this chapter is the "Science Primer" on the NCBI website (see `http://www.ncbi.nlm.nih.gov/About/primer/`). For the engineer or computer scientist who would like to learn about cellular biology, the textbook by Tozeren and Byers (2003) provides a detailed yet approachable description of the field. A remarkably useful and humorous introduction to genetics can be found in the book by Gonick and Wheelis (1983). When learning a new field, one of the most important tasks is to learn the terms used in that field. There are numerous dictionaries of biological and genetic terms, including one by King and Stansfield (1990). Another excellent resource for finding definitions of terms is at `http://www.wikipedia.org`. Finally, for the person who wants to really dig into the subject matter, there are several excellent textbooks (Berg *et al.*, 2002; Watson *et al.*, 2003; Alberts *et al.*, 2002).

The first report of isolation of phage λ was by Esther Lederberg in 1951 (Lederberg, 1951). This discovery marked the beginning of extensive studies of this virus and lysogeny. An excellent history of this work is presented in Gottesman and Weisberg (2004). An important step in this work was the discovery that induction could be activated by UV light (Lwoff and Gutmann, 1950). Lwoff, Jacob, and Monod realized that development of phage λ is essentially controlled by a genetic switch in which genes turn on and off to determine whether the virus proceeds down the lysis or lysogeny pathway (Jacob and Monod, 1961). One of the pioneers in phage λ research, Mark Ptashne, published an excellent description of this genetic switch (Ptashne, 1992). The contents of Section 1.7 are heavily inspired by this text. Finally, the reaction-based model for the phage λ decision circuit as well as its parameters are based on the one proposed in Arkin *et al.* (1998).

Problems

1.1   Consider the first part of an enzymatic reaction:

$$E + S \overset{k_1}{\underset{k_{-1}}{\rightleftarrows}} ES$$

Assume that k_1 is $0.01\text{sec}^{-1}nM^{-1}$, k_{-1} is 0.1sec^{-1}, $[E]$ is 35 nM, $[S]$ is 100nM, $[ES]$ is 50nM, and $RT = 0.5961$ kcal mol^{-1} (i.e., $T = 300°$K). Determine the change in the Gibb's Free Energy for the forward reaction. Is the forward or reverse reaction favored? Using trial-and-error find the point (the concentrations of $[E]$, $[S]$, and $[ES]$) in which this reaction reaches a steady-state (hint: remember that for every nM that you add to $[ES]$, you must take an equal amount off of $[E]$ and $[S]$).

1.2   Consider the binding reactions to O:

$$O + \text{RNAP} \overset{k_1}{\underset{k_{-1}}{\rightleftarrows}} C_1$$

$$O + R \overset{k_3}{\underset{k_{-3}}{\rightleftarrows}} C_2$$

Assuming that $k_{-1} = k_{-3} = 1.0$ sec^{-1}, $\Delta G° = -12.5$ kcal mol^{-1} for the first reaction, $\Delta G° = -12.0$ kcal mol^{-1} for the second reaction, and $RT = 0.614$ kcal mol^{-1} (i.e., $309°$K), what are the values of k_1 and k_3 (hint: the units should be (nM sec)$^{-1}$, and to get these units you need to multiply your answer by 1×10^{-9}).

1.3   Construct a reaction-based SBML model for the genetic circuit shown below using `iBioSim` or any other tool that includes an SBML editor. Use the parameter values provided and assume that CI dimerizes before acting as a transcription factor.

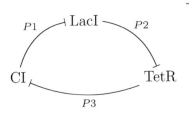

Constant	Value
RNAP$_0$	30 nM
K_d	0.05 M^{-1}
K_o	0.033 M^{-1}
K_r	0.25 M^{-1}
k_o	0.05 sec^{-1}
k_d	0.0075 sec^{-1}
np	10
nc	2

1.4 Construct a reaction-based SBML model for the genetic circuit shown below using `iBioSim` or any other tool that includes an SBML editor. Use the parameter values provided and assume that CI and Cro dimerize before acting as a transcription factor.

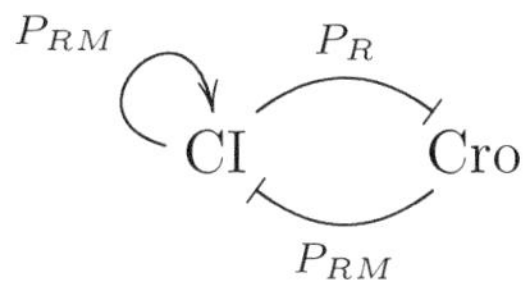

Constant	Value
$RNAP_0$	$30\ nM$
K_d	$0.05\ M^{-1}$
K_o	$0.033\ M^{-1}$
K_a	$0.00001\ M^{-1}$
K_r	$0.25\ M^{-1}$
k_o	$0.05\ sec^{-1}$
k_b	$0.0001\ sec^{-1}$
k_a	$0.25\ sec^{-1}$
k_d	$0.0075\ sec^{-1}$
np	10
na	2
nc	2

1.5 Construct a reaction-based SBML model for the genetic circuit shown below using `iBioSim` or any other tool that includes an SBML editor. Use the parameter values provided and assume that CI dimerizes before acting as a transcription factor.

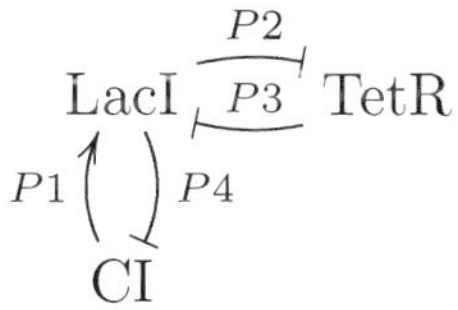

Constant	Value
$RNAP_0$	$30\ nM$
K_d	$0.05\ M^{-1}$
K_o	$0.033\ M^{-1}$
K_a	$0.00001\ M^{-1}$
K_r	$0.25\ M^{-1}$
k_o	$0.05\ sec^{-1}$
k_b	$0.0001\ sec^{-1}$
k_a	$0.25\ sec^{-1}$
k_d	$0.0075\ sec^{-1}$
np	10
na	2
nc	2

1.6   Construct a reaction-based SBML model for the genetic circuit shown below using `iBioSim` or any other tool that includes an SBML editor. Use the parameter values provided and assume that CI dimerizes before acting as a transcription factor.

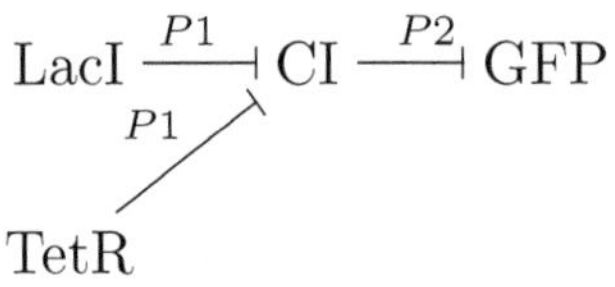

Constant	Value
$RNAP_0$	$30\ nM$
K_d	$0.05\ M^{-1}$
K_o	$0.033\ M^{-1}$
K_r	$0.25\ M^{-1}$
k_o	$0.05\ sec^{-1}$
k_d	$0.0075\ sec^{-1}$
np	10
nc	2

LacI $\xrightarrow{P1}$ CI $\xrightarrow{P2}$ GFP

$\xrightarrow{P1}$

TetR

1.7   Construct a reaction-based SBML model for the portion of the phage λ decision circuit shown in Figure 1.31 which is involved in CI production from P_{RE} and CII production from P_R using `iBioSim` or any other tool that includes an SBML editor.

1.8   Construct a reaction-based SBML model for the entire phage λ decision circuit described in the Appendix using `iBioSim` or any other tool that includes an SBML editor.

1.9   Research literature about a genetic circuit and construct a reaction-based SBML model for this circuit using `iBioSim` or any other tool that includes an SBML editor.

Appendix

This appendix describes the complete chemical reaction genetic circuit model for the phage λ decision circuit. There are a variety of different ways this system can be modeled using chemical reactions, so this is just one possible model. This genetic circuit model includes the five genes (*cI*, *cro*, *cII*, *cIII*, and *N*) and four promoters (P_{RM}, P_R, P_{RE}, and P_L) that are involved in the decision between lysis and lysogeny. A schematic diagram of this model is shown in Figure 1.32. In this diagram, the solid boxes are genes, the hollow boxes are operator sites, the bent arrows are promoters, the diamonds are NUT sites, and the breaks are terminator switches. This model also includes the biochemical reactions involved in dimerization of CI and Cro as well as those involved in the degradation of each protein.

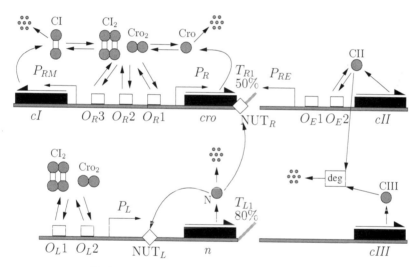

FIGURE 1.32: Phage λ decision circuit.

The reactions shown in Figure 1.33 model the behavior of the P_{RE} promoter. The first reaction represents the binding of RNAP to P_{RE}. This reaction is reversible, and it has an equilibrium constant (ratio of the forward over the reverse rate constant) K_{PRE2} with a value of 0.01 M^{-1}. Note that equilibrium constants are shown with capital letters, and rate constants with lower-case letters. Once RNAP binds to P_{RE}, transcription may begin to produce an mRNA transcript for the *cI* gene. This mRNA transcript can then be translated into the protein CI. The fourth reaction combines both

$$P_{RE} + \text{RNAP} \overset{K_{PRE2}}{\rightleftharpoons} P_{RE} \cdot \text{RNAP}$$

$$P_{RE} + \text{CII} \overset{K_{PRE3}}{\rightleftharpoons} P_{RE} \cdot \text{CII}$$

$$P_{RE} + \text{CII} + \text{RNAP} \overset{K_{PRE4}}{\rightleftharpoons} P_{RE} \cdot \text{CII} \cdot \text{RNAP}$$

$$P_{RE} \cdot \text{RNAP} \overset{k_{PREb}}{\rightarrow} P_{RE} \cdot \text{RNAP} + np\ \text{CI}$$

$$P_{RE} \cdot \text{CII} \cdot \text{RNAP} \overset{k_{PRE}}{\rightarrow} P_{RE} \cdot \text{CII} \cdot \text{RNAP} + np\ \text{CI}$$

Constant	Value	Constant	Value
K_{PRE2}	$0.01\ M^{-1}$	k_{PREb}	$0.00004\ \text{sec}^{-1}$
K_{PRE3}	$0.00726\ M^{-1}$	k_{PRE}	$0.015\ \text{sec}^{-1}$
K_{PRE4}	$0.00161\ M^{-2}$	np	10

FIGURE 1.33: Chemical reaction network model of the promoter P_{RE}.

transcription and translation of this mRNA transcript resulting in np new CI proteins. The value np is the number of proteins produced on average from an mRNA transcript before it is degraded. The rate for this reaction is k_{PREb} with a value of $0.00004\ \text{sec}^{-1}$ which represents the basal rate of *open complex formation* at P_{RE} (i.e., the rate at which transcription is initiated in this configuration). The second reaction represents the binding of the transcription factor CII to an operator site associated with P_{RE}. This complex does not result in transcription. The third reaction represents the binding of CII and RNAP to the operator and promoter site, respectively. The protein CII is an activator meaning that this complex leads to an increased rate of transcription. This can be seen in the fifth reaction which produces CI via transcription and translation. In this case, however, the activated rate is $0.015\ \text{sec}^{-1}$.

Using the modeling method just described requires $(n-1)$ chemical reactions where n is the number of potential configurations of transcription factors and RNAP bound to the operator and promoter sites for a gene. It also requires another m chemical reactions where $m \leq n$ are the number of configurations that lead to transcription. The O_R operator has 40 potential configurations since O_R3 has four possible states (empty, bound to CI_2, bound to Cro_2, and bound to RNAP) and O_R2 and O_R1 may also bind to each transcription factor but must bind to RNAP jointly (total of 10 possibilities). Of these 40 configurations, 13 result in transcription with one resulting in transcription in both directions. This results in a model with 52 reactions.

The first group of reactions shown in Figure 1.34 are those where one molecule is bound to O_R. There are three potential reactions in which CI_2 could bind to each of the three possible operator sites. In this model, this is collapsed into one reaction that indicates that one molecule of CI_2 has bound somewhere to O_R. Similarly, the three reactions for Cro_2 are collapsed into one. Those for RNAP cannot be collapsed because the behavior is different depending on where it binds. Namely, if RNAP binds to O_R3 as in the third reaction, transcription is initiated that results in the production of CI at its basal rate, k_{PRMb}. If, however, RNAP binds to O_R12 (i.e., O_R1 and O_R2), then the transcription initiated produces Cro at a rate of k_{PR}.

$$O_R + CI_2 \xrightleftharpoons{K_{PR2}} O_R \cdot CI_2$$

$$O_R + Cro_2 \xrightleftharpoons{K_{PR5}} O_R \cdot Cro_2$$

$$O_R + RNAP \xrightleftharpoons{K_{PR8}} O_R3 \cdot RNAP$$

$$O_R + RNAP \xrightleftharpoons{K_{PR9}} O_R12 \cdot RNAP$$

$$O_R3 \cdot RNAP \xrightarrow{k_{PRMb}} O_R3 \cdot RNAP + np\ CI$$

$$O_R12 \cdot RNAP \xrightarrow{k_{PR}} O_R12 \cdot RNAP + np\ Cro$$

Constant	Value	Constant	Value
K_{PR2}	$0.2165\ M^{-1}$	K_{PR5}	$0.449\ M^{-1}$
K_{PR8}	$0.1362\ M^{-1}$	K_{PR9}	$0.69422\ M^{-1}$
k_{PRMb}	$0.001\ \sec^{-1}$	k_{PR}	$0.014\ \sec^{-1}$
np	10		

FIGURE 1.34: Model with one species bound to the O_R operator.

The second group of reactions shown in Figure 1.35 are those in which two molecules are bound to O_R. Even using the simplification just described results in nine distinct configurations and 16 reactions for this case. Note that the equilibrium constants for each group are roughly related to the likelihood of each configuration. Equilibrium constants across groups though are incomparable. Two of the most common configurations have either two molecules of CI_2 or one molecule each of CI_2 and Cro_2 bound to O_R which do not lead to transcription. It is also possible to have two molecules of RNAP bound to O_R which leads to the production of both CI and Cro. The most common configuration in this group though is to have a molecule of Cro_2 bound to O_R3 and RNAP bound to the other two sites leading to production of Cro.

The final set of reactions shown in Figure 1.36 are those with three molecules bound to O_R. There are only two common configurations. The first is two molecules of CI_2 are bound (usually to O_R1 and O_R2) and one molecule of Cro_2 is bound. The second common case is RNAP bound to O_R3 and CI_2 bound to the other two sites leading to production of CI at its activated rate of production.

The proteins CI and Cro must dimerize before they can act as transcription factors. The CI and Cro monomers can also degrade. The dimerization and degradation reactions for CI and Cro are shown in Figure 1.37.

$$O_R + 2\text{CI}_2 \overset{K_{PR10}}{\rightleftharpoons} O_R \cdot 2\text{CI}_2$$

$$O_R + 2\text{Cro}_2 \overset{K_{PR13}}{\rightleftharpoons} O_R \cdot 2\text{Cro}_2$$

$$O_R + \text{CI}_2 + \text{Cro}_2 \overset{K_{PR17}}{\rightleftharpoons} O_R \cdot \text{CI}_2 \cdot \text{Cro}_2$$

$$O_R + 2\text{RNAP} \overset{K_{PR16}}{\rightleftharpoons} O_R \cdot 2\text{RNAP}$$

$$O_R + \text{CI}_2 + \text{RNAP} \overset{K_{PR23}}{\rightleftharpoons} O_R \cdot \text{CI}_2 \cdot \text{RNAP}$$

$$O_R + \text{Cro}_2 + \text{RNAP} \overset{K_{PR26}}{\rightleftharpoons} O_R \cdot \text{Cro}_2 \cdot \text{RNAP}$$

$$O_R + \text{RNAP} + \text{CI}_2 \overset{K_{PR24}}{\rightleftharpoons} O_R \cdot \text{RNAP} \cdot \text{CI}_2$$

$$O_R 13 + \text{RNAP} + \text{CI}_2 \overset{K_{PR25}}{\rightleftharpoons} O_R 13 \cdot \text{RNAP} \cdot \text{CI}_2$$

$$O_R + \text{RNAP} + \text{Cro}_2 \overset{K_{PR27}}{\rightleftharpoons} O_R \cdot \text{RNAP} \cdot \text{Cro}_2$$

$$O_R \cdot 2\text{RNAP} \overset{k_{PRMb}}{\rightarrow} O_R \cdot 2\text{RNAP} + np\ \text{CI}$$

$$O_R \cdot 2\text{RNAP} \overset{k_{PR}}{\rightarrow} O_R \cdot 2\text{RNAP} + np\ \text{Cro}$$

$$O_R \cdot \text{CI}_2 \cdot \text{RNAP} \overset{k_{PR}}{\rightarrow} O_R \cdot \text{CI}_2 \cdot \text{RNAP} + np\ \text{Cro}$$

$$O_R \cdot \text{Cro}_2 \cdot \text{RNAP} \overset{k_{PR}}{\rightarrow} O_R \cdot \text{Cro}_2 \cdot \text{RNAP} + np\ \text{Cro}$$

$$O_R \cdot \text{RNAP} \cdot \text{CI}_2 \overset{k_{PRM}}{\rightarrow} O_R \cdot \text{RNAP} \cdot \text{CI}_2 + np\ \text{CI}$$

$$O_R 13 \cdot \text{RNAP} \cdot \text{CI}_2 \overset{k_{PRMb}}{\rightarrow} O_R 13 \cdot \text{RNAP} \cdot \text{CI}_2 + np\ \text{CI}$$

$$O_R \cdot \text{RNAP} \cdot \text{Cro}_2 \overset{k_{PRMb}}{\rightarrow} O_R \cdot \text{RNAP} \cdot \text{Cro}_2 + np\ \text{CI}$$

Constant	Value	Constant	Value
K_{PR10}	$0.06568\ M^{-2}$	K_{PR13}	$0.03342\ M^{-2}$
K_{PR17}	$0.1779\ M^{-2}$	K_{PR16}	$0.09455\ M^{-2}$
K_{PR23}	$0.00967\ M^{-2}$	K_{PR26}	$0.25123\ M^{-2}$
K_{PR24}	$0.0019\ M^{-2}$	K_{PR25}	$0.02569\ M^{-2}$
K_{PR27}	$0.01186\ M^{-2}$	k_{PRMb}	$0.001\ \text{sec}^{-1}$
k_{PR}	$0.014\ \text{sec}^{-1}$	k_{PRM}	$0.011\ \text{sec}^{-1}$
np	10		

FIGURE 1.35: Model with two species bound to the O_R operator.

$$O_R + 3CI_2 \overset{K_{PR29}}{\longleftrightarrow} O_R \cdot 3CI_2$$

$$O_R + 3Cro_2 \overset{K_{PR30}}{\longleftrightarrow} O_R \cdot 3Cro_2$$

$$O_R + 2CI_2 + Cro_2 \overset{K_{PR31}}{\longleftrightarrow} O_R \cdot 2CI_2 \cdot Cro_2$$

$$O_R + CI_2 + 2Cro_2 \overset{K_{PR34}}{\longleftrightarrow} O_R \cdot CI_2 \cdot 2Cro_2$$

$$O_R + RNAP + 2CI_2 \overset{K_{PR37}}{\longleftrightarrow} O_R \cdot RNAP \cdot 2CI_2$$

$$O_R + RNAP + 2Cro_2 \overset{K_{PR38}}{\longleftrightarrow} O_R \cdot RNAP \cdot 2Cro_2$$

$$O_R + RNAP + Cro_2 + CI_2 \overset{K_{PR39}}{\longleftrightarrow} O_R \cdot RNAP \cdot Cro_2 \cdot CI_2$$

$$O_R + RNAP + CI_2 + Cro_2 \overset{K_{PR40}}{\longleftrightarrow} O_R \cdot RNAP \cdot CI_2 \cdot Cro_2$$

$$O_R \cdot RNAP \cdot 2CI_2 \overset{k_{PRM}}{\longrightarrow} O_R \cdot RNAP \cdot 2CI_2 + np\ CI_2$$

$$O_R \cdot RNAP \cdot 2Cro_2 \overset{k_{PRMb}}{\longrightarrow} O_R \cdot RNAP \cdot 2Cro_2 + np\ CI$$

$$O_R \cdot RNAP \cdot Cro_2 \cdot CI_2 \overset{k_{PRMb}}{\longrightarrow} O_R \cdot RNAP \cdot Cro_2 \cdot CI_2 + np\ CI$$

$$O_R \cdot RNAP \cdot CI_2 \cdot Cro_2 \overset{k_{PRM}}{\longrightarrow} O_R \cdot RNAP \cdot CI_2 \cdot Cro_2 + np\ CI$$

Constant	Value	Constant	Value
K_{PR29}	$0.00081\ M^{-3}$	K_{PR30}	$0.00069\ M^{-3}$
K_{PR31}	$0.02133\ M^{-3}$	K_{PR34}	$0.00322\ M^{-3}$
K_{PR37}	$0.0079\ M^{-3}$	K_{PR38}	$0.00026\ M^{-3}$
K_{PR39}	$0.00112\ M^{-3}$	K_{PR40}	$0.00008\ M^{-3}$
k_{PRMb}	$0.001\ \text{sec}^{-1}$	k_{PRM}	$0.011\ \text{sec}^{-1}$
np	10		

FIGURE 1.36: Model with three species bound to the O_R operator.

$$2CI \overset{K_2}{\longleftrightarrow} CI_2$$

$$2Cro \overset{K_5}{\longleftrightarrow} Cro_2$$

$$CI \overset{k_1}{\longrightarrow} ()$$

$$Cro \overset{k_4}{\longrightarrow} ()$$

Constant	Value
K_2	$0.1\ M^{-1}$
K_5	$0.1\ M^{-1}$
k_1	$0.0007\ \text{sec}^{-1}$
k_4	$0.0025\ \text{sec}^{-1}$

FIGURE 1.37: Model for CI and Cro dimerization and degradation.

The chemical reaction network model for the production and degradation of the protein N is shown in Figure 1.38. The model for the promoter P_L is very similar to the ones for the other promoters. This promoter is repressed by either CI_2 or Cro_2.

$$P_L + Cro_2 \xrightleftharpoons{K_{PL2}} P_L \cdot Cro_2$$

$$P_L + CI_2 \xrightleftharpoons{K_{PL4}} P_L \cdot CI_2$$

$$P_L + 2Cro_2 \xrightleftharpoons{K_{PL7}} P_L \cdot 2Cro_2$$

$$P_L + CI_2 + Cro_2 \xrightleftharpoons{K_{PL8}} P_L \cdot CI_2 \cdot Cro_2$$

$$P_L + 2CI_2 \xrightleftharpoons{K_{PL10}} P_L \cdot 2CI_2$$

$$P_L + RNAP \xrightleftharpoons{K_{PL6}} P_L \cdot RNAP$$

$$P_L \cdot RNAP \xrightarrow{k_{PL}} P_L \cdot RNAP + np \text{ N}$$

$$N \xrightarrow{k_7} ()$$

Constant	Value	Constant	Value
K_{PL2}	$0.4132\ M^{-1}$	K_{PL8}	$0.014\ M^{-2}$
K_{PL4}	$0.2025\ M^{-1}$	K_{PL10}	$0.058\ M^{-2}$
K_{PL6}	$0.6942\ M^{-1}$	k_{PL}	$0.011\ \sec^{-1}$
K_{PL7}	$0.0158\ M^{-2}$	k_7	$0.00231\ \sec^{-1}$

FIGURE 1.38: Model for N production and degradation.

The protein CIII is produced on the same transcripts that produce N. In other words, CIII is produced when RNAP binds to P_L. Some transcripts, however, are terminated before they reach the *cIII* gene resulting in a rate of production that is only 20 percent of the full rate. If, however, the anti-terminator, N, is bound to the RNAP at the NUT_L site, then production proceeds at the full rate. This behavior is modeled with the reactions shown in Figure 1.39. The production of the protein CII is modeled in a similar way in Figure 1.40 except that its transcripts originate from P_R.

$$NUT_L + N \xrightleftharpoons{K_{NUT}} NUT_L \cdot N$$

$$P_L \cdot RNAP + NUT_L \xrightarrow{0.2*k_{PL}} P_L \cdot RNAP + NUT_L + np \text{ CIII}$$

$$P_L \cdot RNAP + NUT_L \cdot N \xrightarrow{k_{PL}} P_L \cdot RNAP + NUT_L \cdot N + np \text{ CIII}$$

FIGURE 1.39: Model for CIII production from P_L ($K_{NUT} = 0.2M^{-1}$).

$$\text{NUT}_R i + \text{N} \xrightarrow{K_{\underline{\text{NUT}}}} \text{NUT}_R i \cdot \text{N} \text{ where } i \in \{1, 2, 3, 4\}$$

$$O_R 12 \cdot \text{RNAP} + \text{NUT}_R 1 \xrightarrow{0.5 * k_{PR}} O_R 12 \cdot \text{RNAP} + \text{NUT}_R 1 + np \text{ CII}$$

$$O_R 12 \cdot \text{RNAP} + \text{NUT}_R 1 \cdot \text{N} \xrightarrow{k_{PR}} O_R 12 \cdot \text{RNAP} + \text{NUT}_R 1 \cdot \text{N} + np \text{ CII}$$

$$O_R \cdot 2\text{RNAP} + \text{NUT}_R 2 \xrightarrow{0.5 * k_{PR}} O_R \cdot 2\text{RNAP} + \text{NUT}_R 2 + np \text{ CII}$$

$$O_R \cdot 2\text{RNAP} + \text{NUT}_R 2 \cdot \text{N} \xrightarrow{k_{PR}} O_R \cdot 2\text{RNAP} + \text{NUT}_R 2 \cdot \text{N} + np \text{ CII}$$

$$O_R \cdot \text{CI}_2 \cdot \text{RNAP} + \text{NUT}_R 3 \xrightarrow{0.5 * k_{PR}} O_R \cdot \text{CI}_2 \cdot \text{RNAP} + \text{NUT}_R 3 + np \text{ CII}$$

$$O_R \cdot \text{CI}_2 \cdot \text{RNAP} + \text{NUT}_R 3 \cdot \text{N} \xrightarrow{k_{PR}} O_R \cdot \text{CI}_2 \cdot \text{RNAP} + \text{NUT}_R 3 \cdot \text{N} + np \text{ CII}$$

$$O_R \cdot \text{Cro}_2 \cdot \text{RNAP} + \text{NUT}_R 4 \xrightarrow{0.5 * k_{PR}} O_R \cdot \text{Cro}_2 \cdot \text{RNAP} + \text{NUT}_R 4 + np \text{ CII}$$

$$O_R \cdot \text{Cro}_2 \cdot \text{RNAP} + \text{NUT}_R 4 \cdot \text{N} \xrightarrow{k_{PR}} O_R \cdot \text{Cro}_2 \cdot \text{RNAP} + \text{NUT}_R 4 \cdot \text{N} + np \text{ CII}$$

FIGURE 1.40: Model for CII production from P_R ($K_{NUT} = 0.2 M^{-1}$).

The last set of reactions are for the degradation of CII and CIII. As mentioned earlier, CIII protects CII from degradation. The exact mechanism for this is unknown. One potential mechanism is that CIII may bind to the same proteases (enzymes that degrade proteins) that bind to CII and could thus shield CII from these proteases. This can be modeled as shown in Figure 1.41.

	Constant	Value
$\text{CII} + \text{P1} \xrightarrow{K_8} \text{CII} \cdot \text{P1}$	K_8	$1.0 \ M^{-1}$
$\text{CII} \cdot \text{P1} \xrightarrow{k_{10}} \text{P1}$	k_{10}	$0.002 \ \text{sec}^{-1}$
$\text{CIII} + \text{P1} \xrightarrow{K_{11}} \text{CIII} \cdot \text{P1}$	K_{11}	$10.0 \ M^{-1}$
$\text{CIII} \cdot \text{P1} \xrightarrow{k_{13}} \text{P1}$	k_{13}	$0.0001 \ \text{sec}^{-1}$
$\text{CII} + \text{P2} \xrightarrow{K_{14}} \text{CII} \cdot \text{P2}$	K_{14}	$0.00385 \ M^{-1}$
$\text{CII} \cdot \text{P2} \xrightarrow{k_{16}} \text{P2}$	k_{16}	$0.6 \ \text{sec}^{-1}$
$\text{CIII} + \text{P2} \xrightarrow{K_{17}} \text{CIII} \cdot \text{P2}$	K_{17}	$1.0 \ M^{-1}$
$\text{CIII} \cdot \text{P2} \xrightarrow{k_{19}} \text{P2}$	k_{19}	$0.001 \ \text{sec}^{-1}$
	P1	$35 \ nM$
	P2	$140 \ nM$

FIGURE 1.41: Model for CII and CIII degradation.

Chapter 2

Learning Models

The sciences do not try to explain, they hardly even try to interpret, they mainly make models. By a model is meant a mathematical construct which, with the addition of certain verbal interpretations, describes observed phenomena. The justification of such a mathematical construct is solely and precisely that it is expected to work.

—Johann Von Neumann

There are many methods for predicting the future. For example, you can read horoscopes, tea leaves, tarot cards, or crystal balls. Collectively, these methods are known as "nutty methods." Or you can put well-researched facts into sophisticated computer models, more commonly referred to as "a complete waste of time."

—Scott Adams

Models are to be used, not believed.

—Henri Theil

The first step of the engineering approach is to use the data produced by experiments to construct mathematical models for the systems of interest. With the advent of high-throughput technologies such as *DNA microarrays*, it is now possible to measure expression levels of thousands of mRNA targets simultaneously. Given this experimental data and an abstract class of potential models for genetic circuit configurations, how can one decide the most likely circuit that generated this data?

This chapter begins by briefly describing modern experimental methods in Section 2.1 and then a model for representing experimental data in Section 2.2. Next, Section 2.3 describes *cluster analysis* which groups genes together that seem to be expressed at the same time and are potentially related in function. Clustering algorithms, however, do not indicate which genes activate or repress other genes. Therefore, Section 2.4 presents learning methods for *Bayesian networks* which can potentially determine how genes interact. Bayesian methods, however, have trouble learning causal relationships, so Section 2.5 presents a method for learning a *causal network*. All of these methods must deal both with the limited size of the data sets and the fact that this data tends be very noisy. Gathering more experimental data, however, can be expensive and time consuming. Therefore, Section 2.6 briefly describes how experiments can be designed to improve learning results.

2.1　Experimental Methods

Advances in molecular biology and genomic technologies have led to an explosion in the amount of biological data being generated by the scientific community. This section briefly describes a few of the experimental methods being used to learn the dynamic behavior of genetic circuits.

One of the most direct ways to see what is happening within a genetic circuit is to add a *reporter gene* such as that for *green fluorescent protein* (GFP). This protein comes from a jellyfish that fluoresces green when exposed to blue light. Since GFP is typically not harmful to a cell, it allows the experimenter to observe the workings of a genetic circuit in a living cell. The downside though is that there is only a small number of colors available making it impossible to observe more than a few genes at a time. An example is shown in Figure 2.1 in which both a green and red fluorescent protein are used as reporters controlled by promoters with the same regulatory sequences. In this example, the cells should actually glow yellow when both green and red fluorescent proteins are produced simultaneously. The fact that many cells glow green or red indicates that noise in gene expression is important.

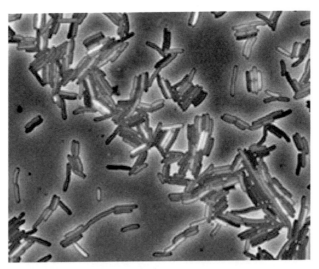

FIGURE 2.1:　This image shows individual cells of the bacterium Escherichia coli expressing red and green fluorescent proteins under the control of identical regulatory sequences. If gene expression were deterministic, all cells would express equal amounts of red and green proteins, and hence appear yellow. The appearance of redder and greener cells in the population indicates the importance of intrinsic randomness, or "noise," in gene expression (image courtesy of Michael B. Elowitz).

In 1995, DNA microarrays were developed that make it possible to monitor the expression levels of thousands of genes simultaneously. DNA microarrays are chips made of glass, plastic, or silicon to which an array of thousands or tens of thousands of single-stranded *complementary DNA* (cDNA) probes have been attached. Each probe represents a segment of a gene about 20 bases in length from the genome of interest. The process for a typical DNA microarray experiment is shown in Figure 2.2 in which differential gene expression is measured between two samples. One sample may be taken from normal cells while the other is taken from tumor cells. Alternatively, one sample could be taken from the start of an experiment monitoring a cells response to some stress factor while the second sample could be taken five minutes later. The first step in this experiment is to isolate the mRNA from the two samples. Next, the mRNA is treated with reverse transcriptase which transcribes the mRNA into single-stranded DNA. At this point, the mRNA from the first sample is labeled with a green fluorescent tag while the second sample is given a red fluorescent tag. The samples are then mixed and hybridized to the complementary probes on the DNA microarray. The microarray is then scanned to determine the color of each spot. In this example, a green spot indicates that a gene is only expressed in normal cells, a red spot indicates that it is only expressed in tumor cells, and a yellow spot indicates that it is present in both. Note that the intensities can also be measured to indicate how much more or less a gene is being expressed in one sample over the other.

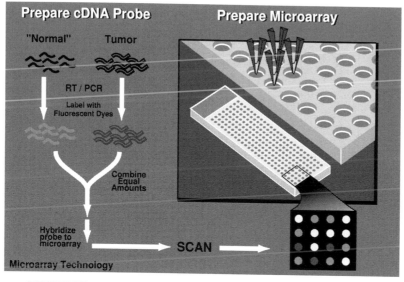

FIGURE 2.2:   Procedure for a DNA microarray experiment (courtesy: National Human Genome Research Institute).

Microarrays can be used to measure gene expression levels which are correlated with the levels of mRNA present. During an experiment, several microarray measurements can be taken at different times to generate *time series experimental data*. This data allows the experimenter to observe the changes in the expression patterns over time in response to a stimulus. While such experiments do allow substantially more genes to be observed than with fluorescent proteins, the cells have to be destroyed to extract their mRNA. Therefore, the time series experimental data collected is for a population rather than an individual cell. Another disadvantage of microarrays is that expression level does not always precisely track the amount of a protein in a cell as this is also dependent on other factors such as degradation rates.

While there are about 25,000 genes in the human genome, there may be more than 500,000 proteins in the human proteome. This observation means that many transcripts must result in many different proteins through alternative splicing and other post-transcriptional modifications, further complicating the ability of DNA microarray results to give an accurate estimate of the quantity of each protein inside a cell. Therefore, proteomic experimental techniques have been developed with the goal of determining the full set of proteins produced in a cell in a given situation.

There are three main steps to proteomic methods. The first step is that the proteins must be separated from each other. This step typically uses *two-dimensional gel electrophoresis*. In one dimension, the proteins are drawn through the gel using an electric field. The distance they travel is related to their molecular weight. In the second dimension, a pH gradient is applied to the gel, and the proteins move until they reach their *isoelectric point*, the pH where the proteins have no net charge. An example result of a 2D gel electrophoresis experiment is shown in Figure 2.3(a). The second step is that each separated protein is analyzed using *mass spectrometry*. The proteins are typically first treated with a protease to break it down into a smaller *peptide*. This peptide is then crystallized on a matrix, ionized with a laser, and the resulting ions are shot towards a detector as shown in Figure 2.3(b). The higher the mass of an ion, the farther it travels. Therefore, the number of ions collected from a variety of collectors provides the molecular weights of the components of the peptide. An example is shown in Figure 2.3(c). This result yields the exact chemical composition of the peptide with a very high precision. From this information, the amino acid sequence and thus the DNA sequence that produced this peptide can be found. In the third step, the genome database for the organism being studied is analyzed to determine which gene produced this protein.

Using data from the experimental methods just described, it is possible to infer the connectivity of the genetic circuit that generated this data using methods described later in this chapter. However, there are also experimental techniques that can be employed to determine which proteins interact as well as which proteins serve as transcription factors for particular genes. Let us consider two techniques here: *two-hybrid screening* and *ChIP-on-chip* which is also known as *genome-wide location analysis*.

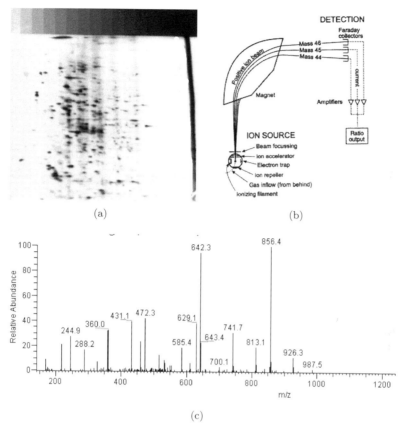

(a)

(b)

(c)

FIGURE 2.3: (a) A 2D gel electrophoresis image from a human leukemia study (courtesy of http://www.lecb.ncifcrf.gov/2DgelDataSets/). (b) Mass spectrometry schematic (courtesy of United States Geological Survey). (c) A protein mass spectrometry result (courtesy of http://en.wikipedia.org/wiki/Protein_mass_spectrometry).

Two-hybrid screening is used to determine if two proteins interact (i.e., bind together). The basic idea of two-hybrid screening is shown in Figure 2.4. A eukaryotic transcription factor is typically composed of two modular domains. The *binding domain* (BD) of the transcription factor is responsible for binding to the operator site while the *activating domain* (AD) is responsible for activating transcription. Two-hybrid screening makes use of the fact that most eukaryotic transcription factors can still function even if their two domains are split apart but are still brought close together. Yeast two-hybrid screening with the GAL4 transcription factor is one of the most commonly used. The reporter gene can either be a fluorescent protein as described earlier, a protein necessary for cell survival (i.e., positive selection), or a protein which kills the cell (i.e., negative selection). In this approach, the *bait* (i.e., the protein being studied) is fused to the BD of GAL4 while the *library protein* is bound to the AD of GAL4. If the bait and library protein do not interact, then the reporter gene is not activated. However, if they do interact, then the reporter gene is activated.

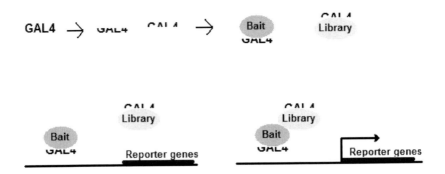

FIGURE 2.4:   Method for two-hybrid screening. The binding and activating domain of the transcription factor GAL4 is split. The binding domain is fused to the bait while the activating domain is fused to the library protein. If the reporter gene is active, this indicates that the bait and library protein interact (courtesy of
`http://en.wikipedia.org/wiki/Two-hybrid_screening`).

While a one-hybrid screening approach can be used to find protein-DNA interactions, ChIP-on-chip, or genome-wide location analysis, is proving to be more effective.   ChIP-on-chip stands for *chromatin immunoprecipitation-on-chip* and the second chip refers to the use of a microarray chip in a later stage of the processing. The goal of this method is to determine which specific DNA sequences can bind to a protein. These DNA binding sites are typically locations where transcription is activated or repressed, and the protein that binds to it is a transcription factor. The ChIP-on-chip process begins by treating cells with a chemical that causes proteins bound to the DNA to be fixated to it. The cells are then lysed to release the DNA which is then fragmented. Next, immunoprecipitation using a specific antibody is employed to separate out the DNA fragments bound to a particular protein. At this point, hydrolysis is applied to make the DNA single-stranded, and these DNA fragments are amplified. They are then labeled and hybridized to a microarray. Finally, the microarray data is analyzed to determine the binding sites for a given transcription factor.

2.2 Experimental Data

The methods described is this chapter begin with a set of experimental data, E. Each experimental data point in E can be expressed using a 3-tuple $\langle e, \tau, \nu \rangle$ in which:

- $e \in \mathbb{N}$ is a natural number representing the experiment number,

- $\tau \in \mathbb{R}$ is the time at which the species values were measured, and

- $\nu \in (\mathbb{R} \cup \{L\} \cup \{H\} \cup \{-\})^{|S|}$ is the state of each species $s \in S$.

The data values of L and H represent that a species is mutated low or high, respectively, in a data point. The data value of '$-$' represents that a species has an unknown value in a data point. The notation $\nu(s)$ denotes the value of species s for that data point. The notation $|E|$ represents the total number of data points within E. Time series experimental data includes multiple time points for the same experiment number. An example set of time series experimental data is shown in Figure 2.5(a). One example data point is $\langle 1, 10, \nu \rangle$ with $\nu(CIII)$ equal to 16. The complete set of data includes 20 time series experiments that include 21 data points that are separated by 5 minutes each. Therefore, $|E|$ is equal to 420.

Before analyzing a set of data, it is typically discretized into a small number of *bins*, n. Each bin j is composed of a range of values $\Phi_j(s) = [\phi_j(s), \phi_{j+1}(s))$ where $j = 0$ to $n - 1$. The ranges are defined by a group of *levels* for each species s of the form $\phi(s) = \langle \phi_0(s), \ldots, \phi_n(s) \rangle$ where $\phi_0(s) = 0$ and $\phi_n(s) = \infty$. A *bin assignment*, $b \in \{0, .., n - 1, *\}^{|S|}$, assigns each $s \in S$ to a particular bin. The notation $b(s)$ returns the bin assignment for species s in b. A bin assignment of '$*$' for a species s indicates that there is no bin assignment specified for s. In this case, the range for a bin of '$*$' is $\Phi_*(s) = [0, \infty)$. A bin assignment that includes $*$'s is called a *partial bin assignment*.

The levels that define the bins can either be provided by the user or determined automatically. For example, an *equal spacing level assignment* can be used in which the range of the data values is determined, and levels are assigned to divide the range into equal size bins. For example, consider the synthetic time series experimental data for the phage λ decision circuit shown in Figure 2.5(a). Figure 2.5(b) shows a histogram that shows the number of times a given expression level for the gene *cIII* is seen in this data. This figure also shows its equal spacing level assignment which is $\phi(cIII) = \langle 0, 33, 67, \infty \rangle$. Figure 2.5(d) is the result of using this level assignment to encode the experimental data shown in Figure 2.5(a). Another method is to use an *equal data level assignment* in which the data is divided as equally as possible between the bins. Figure 2.5(c) shows an equal data level assignment which is $\phi(cIII) = \langle 0, 7, 31, \infty \rangle$, and the result of encoding the data using this level assignment is shown in Figure 2.5(e).

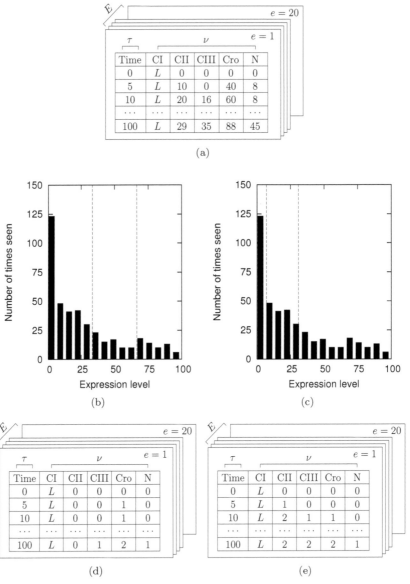

FIGURE 2.5: Time series experimental data (courtesy of Barker (2007)). (a) A set of data. (b) A histogram showing the number of times a given expression level for the gene *cIII* is seen in the experimental data with an equal spacing level assignment of $\phi(cIII) = \langle 0, 33, 67, \infty \rangle$. (c) The same histogram with an equal data level assignment of $\phi(cIII) = \langle 0, 7, 31, \infty \rangle$. (d) Data encoded using the equal spacing level assignment. (e) Data encoded using the equal data level assignment.

2.3 Cluster Analysis

Cluster analysis is often applied to microarray data that have been collected over a variety of conditions or a series of time points. This technique assumes that genes that are active at the same time are likely to be involved in the same regulatory process. It also assumes that genes are grouped and within a group the genes would produce the same expression profile. However, due to noise and other uncertainties during the measurement, these groupings are no longer clear. The goal of cluster analysis is to determine the original groupings of the genes. Cluster analysis must assume that there exists a method to determine the pairwise distance between the expression profiles of any two genes, thus telling the algorithm how similar their expression profiles are. Many algorithms have been proposed to perform clustering. This section only briefly describes two common ones: *K-means* and *agglomerative hierarchical clustering*.

The K-means algorithm partitions N genes into K clusters. This algorithm begins with K initial clusters of the genes either determined by the user or by random. For each cluster, it computes its *centroid* which is essentially the average expression profile of the genes in this cluster. It then reassigns each gene to the cluster with the centroid that is closest to the expression pattern of the gene. At this point, the centroids are recalculated and the process repeats until there is little or no change. Figure 2.6 shows an example in which the squares are the data being clustered while the circles are the centroids. Figure 2.6(a) shows the initial randomly placed centroids, and the cluster assignments using these centroids is shown in Figure 2.6(b). The centroids are then moved to the center of the clusters as shown in Figure 2.6(c), and the cluster assignments are redetermined as shown in Figure 2.6(d).

Figure 2.7 shows microarray gene expression data for the genes in phage λ collected at 20 minutes, 40 minutes, and 60 minutes after an induction event. The K-means algorithm results in the clustering shown in Figure 2.7(a) in which the genes have been partitioned into four clusters as follows:

$$\pi_{K1} = \{cI\}$$
$$\pi_{K2} = \{B, C, U, V, H, M, L, K, I, J\}$$
$$\pi_{K3} = \{A, D, E, Z, G, T, S, R\}$$
$$\pi_{K4} = \{cro, cII, cIII, N, Q, xis, int, O\}$$

These groupings do appear to be biologically significant. The π_{K1} cluster includes only cI which is a lysogeny gene. The π_{K4} cluster includes the early lysis genes. Finally, the π_{K2} and π_{K3} clusters include the later lysis genes.

Another approach is to use *agglomerative hierarchical clustering*. This approach begins with N clusters each containing a single gene. The algorithm proceeds to combine the two clusters that are the smallest distance apart

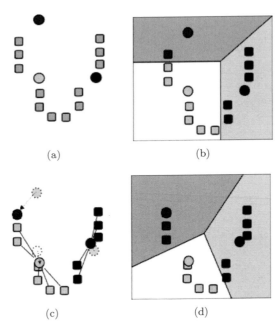

FIGURE 2.6: K-means example (courtesy of http://en.wikipedia.org/wiki/K-means_algorithm). (a) Squares are data being clustered and circles are initial random centroids. (b) Cluster assignment for initial centroids. (c) Centroids are moved to center of initial clusters. (d) Cluster assignment for the new centroids.

where distance is defined as the distance between their average expression profiles. This process continues for $N - 1$ steps at which point all the genes are merged together into a hierarchical tree. A hierarchical clustering for the phage λ microarray data is shown in Figure 2.7(b). The hierarchical clustering algorithm first combines genes A and B. This cluster is then combined with gene D followed by being combined with gene C. Similarly, genes Z and G combine into one cluster while genes E and T combine into another. Then, these two cluster combine with each other and continue to be combined with the $\{A, B, D, C\}$ cluster. This process continues until all genes are combined except cI which is the last gene combined with the others. The result is the hierarchical tree shown to the left of Figure 2.7(b). Using this tree, one possible partition into four clusters is as follows:

$$\pi_{H1} = \{cI\}$$
$$\pi_{H2} = \{cro, cII, cIII, N, Q, xis, int, O\}$$
$$\pi_{H3} = \{S, R\}$$
$$\pi_{H4} = \{A, D, E, Z, G, T, B, C, U, V, H, M, L, K, I, J\}$$

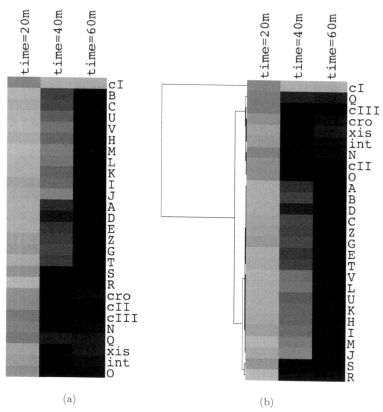

(a) (b)

FIGURE 2.7: Clustering examples of microarray gene expression data for phage λ during induction (generated using Cluster and Java TreeView using data from Osterhout *et al.* (2007)) (a) Clustering using K-means. (b) Clustering using agglomerative hierarchical clustering.

The π_{H1} and π_{H2} clusters are equivalent to the π_{K1} and π_{K4} clusters, respectively. However, the other two clusters are different. This is one challenge with clustering in that different clustering algorithms can yield quite different results. Grouping genes S and R together though does make sense as these two genes work together to lyse the cell.

The major limitation of clustering is that it only results in potential gene relationships, but it does not say how the genes and their protein products interact. To determine this, other methods are needed which is the subject of the remainder of this chapter.

2.4 Learning Bayesian Networks

Given the expression data, E, learning techniques allow one to infer the genetic circuit that best matches E. Bayesian networks are a promising tool for learning the best network from gene expression data. A Bayesian network is a representation of a *joint probability distribution* (PDF). It is represented using a directed acyclic graph, G, whose vertices correspond to random variables, $X_1, \ldots, X_n$, for the expression level of each gene. The connections in this graph represent the dependencies between the random variables.

Consider $P(X, Y)$ to be a joint PDF over two variables X and Y. If $P(X, Y) = P(X)P(Y)$ for all values of X and Y, then X and Y are *independent*. This can also be stated in terms of conditional probability (i.e., $P(X|Y) = P(X)$) in that knowing the value of one variable does not give any information about the other variable. If this is not true, then X and Y are *dependent*. When X and Y are dependent, knowledge of the value of Y provides information about the value of X. It is important to note that when two variables are dependent that they are correlated, but the reverse is not necessarily true. When X and Y are dependent, this is represented in the Bayesian network by an arc between them. If the arc is directed from X to Y, then X is said to be the *parent* of Y.

Associated with each variable X_i is a conditional probability distribution, θ, given its parents in G. In other words, graph G encodes the *Markov Assumption* which states that each variable X_i is independent of its non-descendants given its parents in G. Using the Markov assumption, the joint PDF can be decomposed as follows:

$$P(X_1, \ldots, X_n) = \prod_{i=1}^{n} P(X_i | \mathbf{Pa}(X_i)).$$

where $\mathbf{Pa}(X_i)$ denotes the parents of X_i.

A simple Bayesian network is shown in Figure 2.8(a) and includes both the graph structure, G, and the conditional probability, θ, which is associated with each node in G assuming a binary data encoding. For example, $P(B = 0|A = 1) = 0.8$. The joint PDF is $P(A, B, C) = P(A)P(B|A)P(C|B)$. Therefore, using the conditional probabilities given $P(A = 1, B = 0, C = 0) = 0.5 \times 0.8 \times 0.25 = 0.1$. Another example is shown in Figure 2.8(b). In this example, A is a *common cause* of B and D. If A is not measured, then it is known as a *hidden common cause*. The parents of B are $\mathbf{Pa}(B) = \{A, E\}$. The joint PDF for this example is $P(A, B, C, D, E) = P(A)P(B|A, E)P(C|B)P(D|A)P(E)$.

Graph G implies a set of *conditional independence* assumptions, $\mathrm{Ind}(G)$. For example, the graph in Figure 2.8(a) implies that the value of variable A is independent of the value of variable C given the value of variable B (denoted $A \perp\!\!\!\perp C|B$). The graph in Figure 2.8(b) implies the following conditional

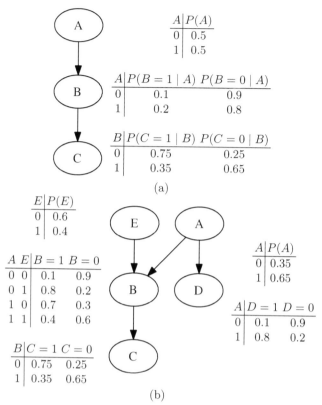

FIGURE 2.8: (a) Simple Bayesian network. (b) Another Bayesian network.

independence relationships: $A \perp\!\!\!\perp C|B$, $A \perp\!\!\!\perp E$, $B \perp\!\!\!\perp D|A$, $C \perp\!\!\!\perp D|A$, $C \perp\!\!\!\perp D|B$, $C \perp\!\!\!\perp E|B$, and $D \perp\!\!\!\perp E$. More than one graph can imply the same set of conditional independence relationships. Graphs G and G' are said to be *equivalent* if $\text{Ind}(G) = \text{Ind}(G')$. Equivalent graphs have the same underlying undirected graph but may disagree on the direction of some edges. For example, if the edges in Figure 2.8(a) are reversed, the resulting graph still implies $A \perp\!\!\!\perp C|B$. For the graph in Figure 2.8(b), reversing any arc other than the one between A and D changes the conditional independence relationships. Bayesian network graphs can be grouped into *equivalence classes* which can be represented by *partially directed acyclic graphs* where edges can be of the following forms: $X \rightarrow Y$, $X \leftarrow Y$, or $X—Y$.

The goal of *Bayesian learning* is to infer the most likely Bayesian network, $\langle G, \theta \rangle$, that may have produced a given a set of experimental data, E (also known as the *training set*). These learning methods utilize a *Bayesian scoring metric* as defined below:

$$P(G|E) = \frac{P(E|G)P(G)}{P(E)}$$

$$\text{Score}(G : E) = \log P(G|E) = \log P(E|G) + \log P(G) + C$$

where $C = -\log P(E)$ is constant, and $P(E|G) = \int P(E|G, \theta)P(\theta|G)d\theta$ is the *marginal likelihood*. The choice of *priors* $P(G)$ and $P(\theta|G)$ influence the score. For example, $P(G)$ may be chosen to penalize graphs with more connections, since a goal is usually to find the simplest graph that explains the data. Given the priors and the experimental data, Bayesian learning attempts to find the structure G that maximizes this score. This problem is NP-hard, so heuristics like greedy random search must be used. For example, beginning with some initial network, a greedy random search would select an edge to add, delete, or reverse. It would then compute this network's score, and if it is better than the previous network, then it would keep the change. This process is repeated until no improvement is found for some number of steps.

The number of Bayesian network graphs is super-exponential in the number of variables, so it is essential to focus this search as much as possible. The *sparse candidate algorithm* identifies small numbers of candidate parents for each gene based on local statistics. One potential pitfall though is that early choices of parents can overly restrict the search space. One possible solution is to adapt the candidate parent sets during the search.

A Bayesian network representation of a genetic circuit can answer questions such as which genes depend on which other genes. In order to learn a Bayesian network from expression data such as that obtained from microarrays, a random variable is associated with the expression level of each gene. Next, a local probability model must be defined for each variable. One possibility is to discretize gene expression into three bins: significantly lower, similar to, or significantly greater than the expression level of the control. Discretizing measured expression levels into just three bins can lose information, but more bins can be used if there is more resolution in the experimental data. The expression level for the control can either be found experimentally or can simply be the average expression level, and whether the difference is significant can be set using a threshold on the ratio between the measured expression level and that of the control.

As an example, let us consider learning a Bayesian network from the synthetic gene expression data for the phage λ decision circuit shown in Table 2.1. This data was generated using stochastic simulation of the model described in the Appendix of Chapter 1, and it represents 20 time series experiments each composed of 22 time points. To simplify the presentation, the data has been discretized using only two values where 0 indicates that the gene is not expressed while a 1 indicates that the gene is expressed. The probability column labeled "Orig" represents the probability of the corresponding gene expression state being found in the data.

TABLE 2.1: Synthetic gene expression data for phage λ

cIII	N	cII	cro	cI	Orig	BN	cIII	N	cII	cro	cI	Orig	BN
		Genes			Probability				Genes			Probability	
0	0	0	0	0	0.05	0.08	1	0	0	0	0	0.00	0.01
0	0	0	0	1	0.18	0.17	1	0	0	0	1	0.02	0.01
0	0	0	1	0	0.06	0.08	1	0	0	1	0	0.01	0.01
0	0	0	1	1	0.10	0.17	1	0	0	1	1	0.00	0.01
0	0	1	0	0	0.00	0.04	1	0	1	0	0	0.01	0.01
0	0	1	0	1	0.04	0.01	1	0	1	0	1	0.01	0.01
0	0	1	1	0	0.02	0.04	1	0	1	1	0	0.01	0.01
0	0	1	1	1	0.02	0.01	1	0	1	1	1	0.00	0.01
0	1	0	0	0	0.01	0.03	1	1	0	0	0	0.01	0.01
0	1	0	0	1	0.05	0.06	1	1	0	0	1	0.01	0.01
0	1	0	1	0	0.05	0.03	1	1	0	1	0	0.01	0.01
0	1	0	1	1	0.02	0.06	1	1	0	1	0	0.00	0.01
0	1	1	0	0	0.03	0.03	1	1	1	0	0	0.20	0.10
0	1	1	0	1	0.00	0.01	1	1	1	0	1	0.02	0.04
0	1	1	1	0	0.03	0.03	1	1	1	1	0	0.02	0.10
0	1	1	1	1	0.00	0.01	1	1	1	1	1	0.00	0.04

One possible Bayesian network representation for the phage λ decision circuit is shown in Figure 2.9. Note that once the graph structure G is selected, the conditional probability distribution θ can be determined directly from the data. For example, $P(cIII = 1 | N = 0)$ is determined by summing the probabilities where both $cIII$ is 1 and N is 0 and dividing by the sum of the probabilities where N is 0. Using this network and the corresponding joint PDF shown at the bottom of the figure, the probability of each expression state can be determined, and the results are shown in the probability column labeled "BN" in Table 2.1. In many cases, the probabilities match fairly well lending support to this network representation. However, in some states such as $\langle 11100 \rangle$ and $\langle 11110 \rangle$, there are significant differences.

Several interesting relationships can be inferred from this network. First, the expression of the gene N increases the chance of finding the gene $cIII$ being expressed. This makes sense since N helps transcripts from P_L proceed past the terminator switch and transcribe $cIII$. Second, the expression of either N or $cIII$ increase the chance of finding cII being expressed. Again, this relationship makes sense due to N's role in helping transcripts from P_R past the terminator to transcribe cII, and CIII's role in protecting CII from degradation. Finally, the expression of cI reduces the chance of cro being expressed which makes sense since CI represses P_R. One implication from this network seems incorrect in that the expression of cII seems to inhibit the expression of cI. The reason for this is that while cII expression is needed to get CI production started, CI actually represses further production of CII. Therefore, the majority of the time in which cI is expressed is when cII is not expressed. Such counter-intuitive results should serve as a reminder that

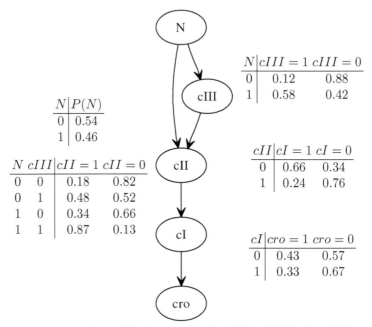

$$P(cIII, N, cII, cro, N) = P(N)P(cIII \mid N)P(cII \mid N, cIII)P(cI \mid cII)P(cro \mid cI)$$

FIGURE 2.9: A Bayesian model for the phage λ decision circuit.

Bayesian networks only show correlations and not causal relationships. This point is further illustrated with the Bayesian network for the phage λ decision circuit shown in Figure 2.10 in which N and $cIII$ have exchanged places. In this network, it would appear that the expression of $cIII$ increases the chance of the expression of N which is not true biologically.

A major difficulty with learning Bayesian networks from gene expression data is that microarray experiments often include data for thousands of genes but typically only a small number of sample points. This limited data size makes it quite likely that there are several networks with similar scores. Therefore, a set of potential networks must be considered. Since genetic circuits often have only a sparse number of connections, one can search for common features in this set of potential networks rather than attempting to find the actual best network. These features can help identify potential causal relationships, but keep in mind that relationships may only be correlative. One feature that can be identified is a *Markov relation*. There is Markov relation between genes X and Y when Y is a member of the minimal set of variables that when controlled for makes X independent of all other variables outside this set. Another type of feature is an *order relation*. There is an ordering relation between genes X and Y if X is an ancestor of Y. In order to estimate *statistical confidence* that the feature is actually a true feature, one can use

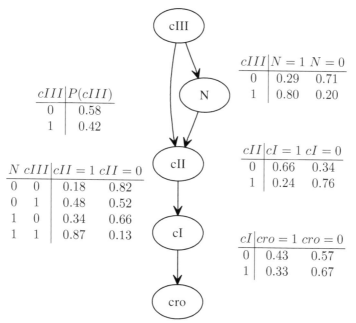

$$P(cIII, N, cII, cro, N) = P(cIII)P(N \mid cIII)P(cII \mid N, cIII)P(cI \mid cII)P(cro \mid cI)$$

FIGURE 2.10: Another Bayesian model for the phage λ decision circuit.

the equation below:

$$\text{confidence}(f) = \frac{1}{m} \sum_{i=1}^{m} f(G_i)$$

where m is the number of potential networks considered, G_i is a potential network, and $f(G_i)$ is 1 if f is a feature of G_i and 0 otherwise. One way to generate many potential networks is to use a *bootstrap* method. A bootstrap method divides the experimental data into several different subsets, and a potential network is generated for each of these subsets of the data.

Clustering approaches can only determine correlations. Bayesian analysis can potentially discover causal relationships and interactions between genes. These models have probabilistic semantics which is well suited to the noisy, stochastic behavior of real biological systems. Bayesian approaches may be useful at extracting features rather than finding a single model. Bayesian analysis can also potentially assist with experimental design. Bayesian networks though have several limitations. As illustrated in the example, they often cannot differentiate between cause and correlation. Another limitation of Bayesian networks is that their graphs must be acyclic. This is a severe limitation since most (if not all) genetic circuits include feedback control.

To address this problem, *dynamic Bayesian networks* (DBN) have been applied to the analysis of temporal expression data. A DBN is essentially a cyclic graph that has been unrolled T times. The nodes in a DBN are random variables, $X_1^{(t)}, \ldots, X_n^{(t)}$ where t is a value between 1 and T. Each node represents the probability of that variable at a given time point. Therefore, an arc from $X_i^{(1)}$ directed to $X_j^{(2)}$ represents an influence between two nodes over time. For a DBN, the joint PDF can be decomposed as follows:

$$P(X_1^{(1)}, \ldots, X_n^{(T)}) = \prod_{t=1}^{T} \prod_{i=1}^{n} P(X_i^{(t)} | \mathbf{Pa}(X_i^{(t)}))$$

Learning a DBN can proceed essentially in the same fashion as learning a Bayesian network except that the data used must be time series data.

An example DBN for the phage λ decision circuit is shown in Figure 2.11. The DBN representation includes several influences that cannot be represented in a Bayesian network. For example, this DBN indicates both that *cI*'s expression level influences *cro*'s expression level as well as that *cI*'s expression level influences *cro*'s expression level. Including both of these influences in a Bayesian network would have created a cycle which is not allowed. Also, auto-regulation can be represented such as the fact that the current expression level of *cI* influences the future expression level of *cI*.

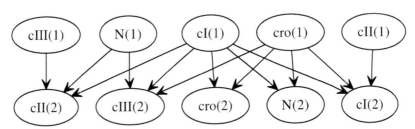

FIGURE 2.11: A DBN for the phage λ decision circuit.

2.5 Learning Causal Networks

A Bayesian network represents correlative relationships, but learning causal relationships is more important. Therefore, one would like to learn a *causal network* in which the parents of a variable are interpreted as its *immediate causes*. The *causal Markov assumption* states that given the values of a variable's immediate causes, its value is independent of all earlier causes. Causal networks model not only distribution of observations but also the effects of *interventions* such as gene mutations. In causal networks, $X \to Y$ and $Y \to X$ are not equivalent. The true causal network can be any network in the equivalence class of Bayesian networks found as described above.

This section presents an efficient method for learning a causal network from time series experimental data. DBN approaches typically must perform an expensive global search in order to find the best fit network. DBN approaches also typically have difficulty learning networks with tight feedback. The method described in this section improves efficiency by utilizing a a *local* analysis to efficiently learn networks with *tight* feedback. In particular, this method considers each gene one at a time. This method determines the likelihood that a gene's expression increases in the next time point given the gene expression levels in the current time point. These likelihoods are then used to determine *influences* between the genes in the genetic circuit. The result is a directed graph representation of the genetic circuit.

The major difficulty which must be dealt with when learning a causal network model of a genetic circuit is that there is a very large number of possible networks to be considered. For example, the five gene genetic circuit for the phage λ decision circuit is shown in Figure 2.12. This is, however, only one out of a potential 8.47×10^{11} genetic circuit models. In general, there are $3^{|S|^2}$ potential models where $|S|$ is the number of species (i.e., genes) in the genetic circuit of interest. This is because there are $|S|^2$ relationships between $|S|$ species, and each pair of species has one of three types of relationships (i.e., repression, activation, or no relationship).

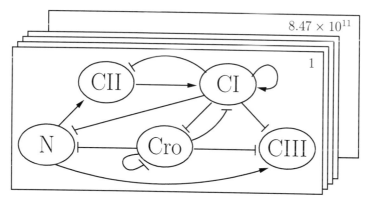

FIGURE 2.12: Possible models for the phage λ decision circuit (courtesy of Barker (2007)).

To reduce complexity, this learning method considers each gene one at a time, and it determines the most likely *influence vector* for that gene. An influence vector for a gene species represents the influence that each species has on the expression of this gene. The possible influences are n for no influence, a for an activating influence, r for a repressing influence, and u for an unknown influence. For example, the influence vector for the gene $cIII$ of the form $i = \langle r, n, n, u, a \rangle$ with species order $\langle$ CI, CII, CIII, Cro, N $\rangle$ specifies that CI represses expression of $cIII$, CII and CIII have no influence on expression of $cIII$, Cro has an unknown relationship, and N activates expression of $cIII$. For convenience, the function $Act(i)$ returns the activators in i, $Rep(i)$ returns the repressors in i, and $Par(i)$ returns all the parents in i. For the example just described, $Act(i) = \{N\}$, $Rep(i) = \{CI\}$, and $Par(i) = \{N, CI\}$.

The goal of the learning method described in this section is to determine an influence vector for each species in S by examining time series experimental data points in E. In particular, this learning method considers pairs of experimental data point that occur in sequence. The set of these data point pairs is defined as follows:

$$\Gamma = \{(\langle e, \tau, \nu \rangle, \langle e', \tau', \nu' \rangle) \mid \langle e, \tau, \nu \rangle \in E \wedge \langle e', \tau', \nu' \rangle \in E$$
$$\wedge (e = e') \wedge (\tau < \tau') \wedge \neg \exists \langle e, \tau'', \nu'' \rangle \in E.(\tau < \tau'') \wedge (\tau'' < \tau')\}$$

For the example in Figure 2.5(a), $(\langle 1, 0, \nu \rangle, \langle 1, 5, \nu' \rangle)$ is a member of Γ while $(\langle 1, 0, \nu \rangle, \langle 2, 5, \nu' \rangle)$, $(\langle 1, 5, \nu \rangle, \langle 1, 0, \nu' \rangle)$, and $(\langle 1, 0, \nu \rangle, \langle 1, 10, \nu' \rangle)$ are not.

This learning method begins with an initial influence vector for each species that contains any *background knowledge* that the modeler may know from previous studies. The learning method is composed of three steps. The first step constructs a set of candidate influence vectors each with a single new influence added to the initial influence vector. The second step attempts to combine vectors in this set to form larger influence vectors (i.e., ones with more parents). Finally, the third step competes the candidate influence vectors to determine the most likely final influence vector.

The goal of the remaining three steps of the learning method is to search for the best influence vector for each species, s, in the genetic circuit of interest. The key to these steps is being able to assign a score to each influence vector considered. The score is determined by finding the probability of gene expression increasing in the time point given that the current time point includes valid data for s and the data for this time point is within one of several partial bin assignments, b. The gene expression for a species is increasing when the value of this species increases between two successive data points.

$$inc(s) = \{(\langle e, \tau, \nu \rangle, \langle e', \tau', \nu' \rangle) \in \Gamma \mid \nu(s) < \nu'(s)\}$$

A pair of data points is valid for a species s when the species is not mutated or unknown in either data point.

$$val(s) = \{(\langle e, \tau, \nu \rangle, \langle e', \tau', \nu' \rangle) \in \Gamma \mid \nu(s) \notin \{L, H, -\} \wedge \nu'(s) \notin \{L, H, -\}\}$$

Finally, a data point pair is included in a partial bin assignment, b, when for every species, s', the value of that species is within the range of the bin specified in b.

$$bin(b) = \{(\langle e, \tau, \nu \rangle, \langle e', \tau', \nu' \rangle) \in \Gamma \mid \forall s' \in S. \nu(s') \in \Phi_{b(s')}(s')\}$$

Utilizing these definitions, it is now possible to define the conditional probability as follows:

$$P(inc(s) \cap val(s) \cap bin(b)) = \frac{|inc(s) \cap val(s) \cap bin(b)|}{|\Gamma|}$$

$$P(val(s) \cap bin(b)) = \frac{|val(s) \cap bin(b)|}{|\Gamma|}$$

$$P(inc(s) \mid val(s) \cap bin(b)) = \frac{P(inc(s) \cap val(s) \cap bin(b))}{P(val(s) \cap bin(b))}$$

$$= \frac{|inc(s) \cap val(s) \cap bin(b)|}{|val(s) \cap bin(b)|}$$

Therefore, calculating this probability amounts to examining each data point in E and counting up the number of valid data points for the species s in the the bin b and determining the number of these in which s is increasing.

In order to score an influence vector, several conditional probabilities are determined for several partial bin assignments b. In particular, every partial bin assignment over the set of parents in the influence vector i (i.e., $Par(i)$) is considered. For example, consider an influence vector of the form $\langle n, r, r, n, n \rangle$ for the specie N from the phage λ decision circuit. This vector implies that both CII and CIII repress the gene N. Assuming that three bins are being used, the bin assignment for CII and CIII can be any value in $\{0, 1, 2\}$ for a total of nine possible bin assignments. The probability of N's expression increasing in each of these partial bin assignments is shown graphically in Figure 2.13 and in tabular form in Table 2.2.

TABLE 2.2: Conditional probabilities, ratios, and votes for an influence vector in which CII and CIII repress gene N (courtesy of Barker (2007))

$\langle CI, CII, CIII, Cro, N \rangle$	$P(N \uparrow \mid b(CII), b(CIII))$	Ratio	Vote
$\langle *, 0, 0, *, * \rangle$	40	base	
$\langle *, 0, 1, *, * \rangle$	49	1.23	v_a
$\langle *, 0, 2, *, * \rangle$	70	1.75	v_a
$\langle *, 1, 0, *, * \rangle$	58	1.45	v_a
$\langle *, 1, 1, *, * \rangle$	42	1.05	v_n
$\langle *, 1, 2, *, * \rangle$	38	0.95	v_n
$\langle *, 2, 0, *, * \rangle$	66	1.65	v_a
$\langle *, 2, 1, *, * \rangle$	38	0.95	v_n
$\langle *, 2, 2, *, * \rangle$	26	0.65	v_f

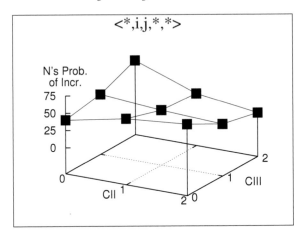

FIGURE 2.13: N's probability of increasing under each possible partial bin assignment for species CII and CIII (courtesy of Barker (2007)).

In order to determine trends, a ratio is formed between two probabilities using two partial bin assignments, b and b', as shown below:

$$\frac{\mathcal{P}(inc(s) \mid val(s) \cap bin(b))}{\mathcal{P}(inc(s) \mid val(s) \cap bin(b'))} = \frac{|inc(s) \cap val(s) \cap bin(b')|}{|val(s) \cap bin(b')|}$$
$$* \frac{|val(s) \cap bin(b)|}{|inc(s) \cap val(s) \cap bin(b)|}$$

The partial bin assignment, b', is called the *base bin assignment*. The base bin assignment is the point of comparison for all other partial bin assignments. The base bin assignment assigns each species in $Act(i)$ to the lowest bin and each species in $Rep(i)$ to the highest bin unless there are more repressors than activators in which case the assignment is reversed. This is defined formally as follows:

$$b'(s) = \begin{cases} * & \text{if } i(s) = \text{`}n\text{'} \\ 0 & \text{if } (i(s) = \text{`}a\text{'} \wedge |Rep(i)| \le |Act(i)|) \vee \\ & \quad (i(s) = \text{`}r\text{'} \wedge |Rep(i)| > |Act(i)|) \\ n-1 & \text{otherwise} \end{cases}$$

For our example, the base bin assignment would be $\langle *, 0, 0, *, * \rangle$, and the probability ratios found using this base bin assignment are shown in Table 2.2.

Each probability ratio is used to determine a vote for (v_f), against (v_a), or neutral to (v_n) the influence vector being evaluated. When there are more activating influences (*i.e.*, $|Rep(i)| \le |Act(i)|$), the votes are determined using the thresholds shown in Figure 2.14(a). When there are more repressing influences (*i.e.*, $|Rep(i)| > |Act(i)|$), the votes are determined using the thresholds shown in Figure 2.14(b). Using a value of 1.15 for T_a and 0.75 for T_r, the votes determined for our example are shown in Table 2.2. The final score for the influence vector is determined using the following equation:

$$score = \frac{v_f - v_a}{v_f + v_a + v_n}$$

A score greater than zero indicates support for the influence vector while a negative score indicates there is not support for the influence vector. This influence vector has a score of -0.375, so the data does not support the influence vector $\langle n, r, r, n, n \rangle$ for gene N.

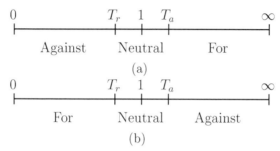

FIGURE 2.14: (a) Thresholds for determining votes when there are more activating influences (*i.e.*, $|Rep(i)| \leq |Act(i)|$). (b) Thresholds for determining votes when there are more repressing influences (*i.e.*, $|Rep(i)| > |Act(i)|$) (courtesy of Barker (2007)).

When scoring an influence vector, i, for species s, the probability of increase can be influenced by the level of s itself. For example, the likelihood of gene expression increasing would usually go down as the overall amount of gene expression goes up. Also, when comparing two influence vectors, i and i', it is useful to control for the species in i', when evaluating the score for i and vice versa. In both of these cases, this method partitions the bin assignments using a control set, G. Now, all bin assignments to species in $Par(i) \cup G$ are considered. The base bin assignment is also redefined such that it agrees with the values in b for each member of G. This is defined formally as follows:

$$b'(s) = \begin{cases} b(s) & \text{if } s \in G \\ * & \text{if } i(s) = \text{`}n\text{'} \\ 0 & \text{if } (i(s) = \text{`}a\text{'} \wedge |Rep(i)| \leq |Act(i)|) \\ & \quad \vee (i(s) = \text{`}r\text{'} \wedge |Rep(i)| > |Act(i)|) \\ n - 1 & \text{otherwise} \end{cases}$$

As an example, again consider the influence vector $\langle n, r, r, n, n \rangle$, but this time let us control for the level of N (i.e., $G = \{N\}$). In other words, probabilities are now calculated for each bin assignment for CII, CIII, and N. However, only those that have the same bin assignment for N are compared. Therefore, there are three probability graphs considered as shown in Figure 2.15. There are now three base bin assignments used to calculate the probability ratios, and there are a total of 24 votes as shown in Table 2.3. The final score is now -0.16, so the data does not support this influence vector.

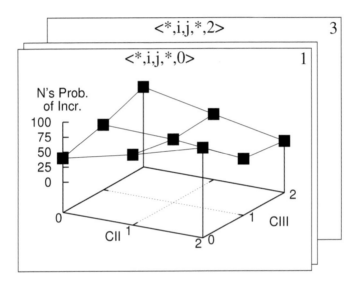

FIGURE 2.15: N's probability of increasing for the vector $\langle n, r, r, n, n \rangle$ with $G = \{N\}$ (courtesy of Barker (2007)).

The goal of a causal network learning method is to efficiently search through the potential influence vectors for each species in S and find the one with the best score. The method described in this section divides this process into three steps. In the first step, a set of influence vectors is created by considering each species in turn as a potential activator or repressor of the gene of interest. As an example, consider an initial influence vector for the gene *cIII* of $\langle u, u, n, u, u \rangle$. Note that self-arcs are currently assumed to be no connection as this method has difficulty in learning auto-regulation. The conditional probabilities and ratios computed for the eight influence vectors constructed from this initial vector are shown in Table 2.4. The votes and scores for each of these influence vectors is shown in Table 2.5. Only the three influence vectors, $\langle r, n, n, n, n \rangle$, $\langle n, n, n, r, n \rangle$, and $\langle n, n, n, n, r \rangle$, have scores high enough to be considered further.

TABLE 2.3: Conditional probabilities, ratios, and votes for the influence vector $\langle n, r, r, n, n \rangle$ with $G = \{N\}$ (courtesy of Barker (2007))

$\langle \text{CI, CII, CIII, Cro, N} \rangle$	$\mathcal{P}(\text{N} \uparrow \mid b(\text{CII}), b(\text{CIII}), b(\text{N}))$	Ratio	Vote
$\langle *, 0, 0, *, 0 \rangle$	40	base	
$\langle *, 0, 1, *, 0 \rangle$	58	1.45	v_a
$\langle *, 0, 2, *, 0 \rangle$	83	2.08	v_a
$\langle *, 1, 0, *, 0 \rangle$	67	1.66	v_a
$\langle *, 1, 1, *, 0 \rangle$	55	1.37	v_a
$\langle *, 1, 2, *, 0 \rangle$	59	1.47	v_a
$\langle *, 2, 0, *, 0 \rangle$	100	2.5	v_a
$\langle *, 2, 1, *, 0 \rangle$	44	1.09	v_n
$\langle *, 2, 2, *, 0 \rangle$	36	0.90	v_n
$\langle *, 0, 0, *, 1 \rangle$	55	base	
$\langle *, 0, 1, *, 1 \rangle$	40	0.72	v_f
$\langle *, 0, 2, *, 1 \rangle$	50	0.90	v_n
$\langle *, 1, 0, *, 1 \rangle$	54	0.98	v_n
$\langle *, 1, 1, *, 1 \rangle$	37	0.67	v_f
$\langle *, 1, 2, *, 1 \rangle$	41	0.75	v_n
$\langle *, 2, 0, *, 1 \rangle$	0	0.00	v_f
$\langle *, 2, 1, *, 1 \rangle$	42	0.76	v_n
$\langle *, 2, 2, *, 1 \rangle$	27	0.49	v_f
$\langle *, 0, 0, *, 2 \rangle$	27	base	
$\langle *, 0, 1, *, 2 \rangle$	22	0.81	v_n
$\langle *, 0, 2, *, 2 \rangle$	50	1.85	v_a
$\langle *, 1, 0, *, 2 \rangle$	30	1.11	v_n
$\langle *, 1, 1, *, 2 \rangle$	28	1.04	v_n
$\langle *, 1, 2, *, 2 \rangle$	28	1.04	v_n
$\langle *, 2, 0, *, 2 \rangle$	100	3.70	v_a
$\langle *, 2, 1, *, 2 \rangle$	30	1.11	v_n
$\langle *, 2, 2, *, 2 \rangle$	24	0.88	v_n

TABLE 2.4: Probabilities and ratios found by the score function with species list $\langle \text{CI, CII, CIII, Cro, N} \rangle$ (courtesy of Barker (2007))

Vector	$P(inc(\text{CIII}) \mid bin(b) \cap val(\text{CIII}))$			Ratios	
	$\Phi_0(\text{CI})$	$\Phi_1(\text{CI})$	$\Phi_2(\text{CI})$	$\frac{\Phi_1(\text{CI})}{\Phi_0(\text{CI})}$	$\frac{\Phi_2(\text{CI})}{\Phi_0(\text{CI})}$
$\langle a, n, n, n, n \rangle$ $\Phi_0(\text{CIII})$	19.0%	1.7%	1.0%	0.09	0.05
$\langle r, n, n, n, n \rangle$ $\Phi_1(\text{CIII})$	17.1%	2.6%	1.2%	0.15	0.07
$\Phi_2(\text{CIII})$	11.6%	2.7%	1.1%	0.23	0.09
	$\Phi_0(\text{CII})$	$\Phi_1(\text{CII})$	$\Phi_2(\text{CII})$	$\frac{\Phi_1(\text{CII})}{\Phi_0(\text{CII})}$	$\frac{\Phi_2(\text{CII})}{\Phi_0(\text{CII})}$
$\langle n, a, n, n, n \rangle$ $\Phi_0(\text{CIII})$	3.1%	13.7%	–	4.32	–
$\langle n, r, n, n, n \rangle$ $\Phi_1(\text{CIII})$	4.4%	7.4%	12.6%	1.65	2.83
$\Phi_2(\text{CIII})$	19.4%	5.5%	6.8%	0.28	0.35
	$\Phi_0(\text{Cro})$	$\Phi_1(\text{Cro})$	$\Phi_2(\text{Cro})$	$\frac{\Phi_1(\text{Cro})}{\Phi_0(\text{Cro})}$	$\frac{\Phi_2(\text{Cro})}{\Phi_0(\text{Cro})}$
$\langle n, n, n, a, n \rangle$ $\Phi_0(\text{CIII})$	11.5%	1.8%	1.5%	0.16	0.13
$\langle n, n, n, r, n \rangle$ $\Phi_1(\text{CIII})$	14.2%	4.7%	3.1%	0.33	0.23
$\Phi_2(\text{CIII})$	9.7%	5.0%	4.2%	0.52	0.43
	$\Phi_0(\text{N})$	$\Phi_1(\text{N})$	$\Phi_2(\text{N})$	$\frac{\Phi_1(\text{N})}{\Phi_0(\text{N})}$	$\frac{\Phi_2(\text{N})}{\Phi_0(\text{N})}$
$\langle n, n, n, n, a \rangle$ $\Phi_0(\text{CIII})$	5.4%	2.9%	3.7%	0.53	0.68
$\langle n, n, n, n, r \rangle$ $\Phi_1(\text{CIII})$	9.3%	7.2%	6.7%	0.78	0.72
$\Phi_2(\text{CIII})$	8.6%	6.4%	6.1%	0.75	0.71

TABLE 2.5: Votes and scores found by the Score function from the probabilities in Table 2.4 for influence vectors containing only a single influence on gene *cIII* in the phage λ decision circuit (courtesy of Barker (2007))

$\langle$CI, CII, CIII, Cro, N$\rangle$	v_f	v_a	v_n	Score
$\langle a, n, n, n, n \rangle$	0	6	0	-1.00
$\langle r, n, n, n, n \rangle$	6	0	0	1.00
$\langle n, a, n, n, n \rangle$	3	2	0	0.20
$\langle n, r, n, n, n \rangle$	2	3	0	-0.20
$\langle n, n, n, a, n \rangle$	0	6	0	-1.00
$\langle n, n, n, r, n \rangle$	6	0	0	1.00
$\langle n, n, n, n, a \rangle$	0	5	1	-0.83
$\langle n, n, n, n, r \rangle$	5	0	1	0.83

It is, of course, possible that multiple species can influence the production from a single gene. Therefore, the next step considers influence vectors found in the previous step two at a time to determine if they can be merged to form a larger influence vector with a better score. When the merged influence vector has a better score than both of the individual influence vectors forming the merge, it is kept. At the end of this step, each of the influence vectors contained in a larger influence vector is removed. For example, the probabilities found for an influence vector that contains both CI and Cro are shown in Figure 2.16. This figure shows that both CI and Cro appear to have a repressive affect on the *cIII* gene which is supported by all 24 votes. Therefore, this influence vector is at least as good as the two individual vectors, so it is retained. All other possible combinations are also considered as shown in Table 2.6, but no other combination is found to be better than the individual vectors. After removing the subset vectors, the resulting influence vectors are: $\langle n, n, n, n, r \rangle$ and $\langle r, n, n, r, n \rangle$.

TABLE 2.6: Votes and scores found when combining influence vectors on gene *cIII* in the phage λ decision circuit (courtesy of Barker (2007))

$\langle$CI, CII, CIII, Cro, N$\rangle$	v_f	v_a	v_n	Score
$\langle r, n, n, n, n \rangle$	6	0	0	1.0
$\langle n, n, n, r, n \rangle$	6	0	0	1.0
$\langle n, n, n, n, r \rangle$	5	0	1	0.833
$\langle r, n, n, r, n \rangle$	24	0	0	1.0
$\langle r, n, n, n, r \rangle$	21	0	3	0.875
$\langle n, n, n, r, r \rangle$	23	0	1	0.958
$\langle r, n, n, r, r \rangle$	71	1	6	0.897

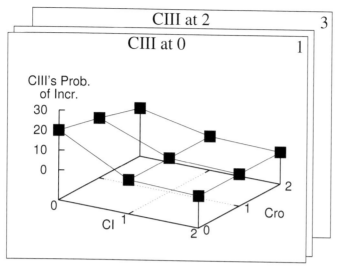

FIGURE 2.16: CIII's probability of increasing with the influence vector $i = \langle r, n, n, r, n \rangle$, where the species order is $\langle \text{CI}, \text{CII}, \text{CIII}, \text{Cro}, \text{N} \rangle$ and $G = \{\text{CIII}\}$ (courtesy of Barker (2007)).

The final step is to compete the remaining influence vectors to determine the best one. Two influence vectors, i and i', are competed at a time. In order to reduce the effect of correlation, when determining the score for i, those species found in $Par(i')$ but not found in $Par(i)$ are added to the control set, G. The reverse is done when finding the score for i'. This has a similar effect as if one generates data using a mutational experiment. After determining the two scores, the influence vector with the lowest score is discarded. Figure 2.17 shows the probabilities that are found when scoring the influence vector that includes N as a repressor while competing it with the one which has Cro and CI. Separate ratios are determined for each encoded value of Cro, CI, and CIII. Similarly, the probabilities found when scoring the influence vector with CI and Cro are shown in Figure 2.18 in which separate ratios are found for each encoded value of N and CIII. The resulting votes and scores are shown in Table 2.7. The influence vector $\langle r, n, n, r, n \rangle$ has the higher score, so the learning method has determined that CI and Cro repress the gene *cIII*.

TABLE 2.7: Scores found when competing potential influence vectors for gene *cIII* (courtesy of Barker (2007))

$\langle \text{CI}, \text{CII}, \text{CIII}, \text{Cro}, \text{N} \rangle$	$votes_f$	$votes_a$	$votes_n$	Score
$\langle n, n, n, n, r \rangle$	12	27	15	0.27
$\langle r, n, n, r, n \rangle$	71	0	1	0.98

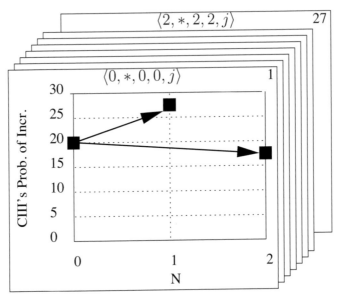

FIGURE 2.17: CIII's probability of increasing with the influence vector $i = \langle n, n, n, n, r \rangle$, where the species order is $\langle CI, CII, CIII, Cro, N \rangle$, and $G = \{CI, CIII, Cro\}$ (courtesy of Barker (2007)).

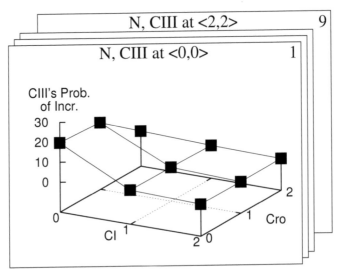

FIGURE 2.18: CIII's probability of increasing with the influence vector $i = \langle r, n, n, r, n \rangle$, where the species order is $\langle CI, CII, CIII, Cro, N \rangle$, and $G = \{CIII, N\}$ (courtesy of Barker (2007)).

Applying the learning method just described to the other species results in the final genetic circuit model shown in Figure 2.19. This model includes 7 of the 12 arcs in the actual genetic circuit. Two of the arcs that are not found are the activation of *cII* and *cIII* by N which is not actually transcriptional regulation as N promotes synthesis of CII and CIII by binding to RNAP so that it can pass over the terminator switches. Two more arcs that are not found are self-arcs that this method is not capable of finding. It should also be noted that no false arcs are reported for this example.

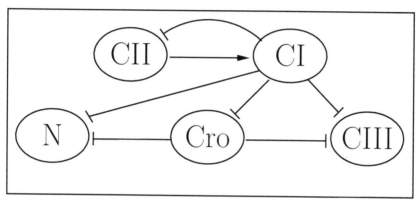

FIGURE 2.19: The final genetic circuit learned for the phage λ decision circuit. All arcs reported are correct. There are, however, 2 activation arcs, 1 repression arc, and the 2 self-arcs that are not found (courtesy of Barker (2007)).

2.6 Experimental Design

When learning a genetic circuit model, many potential models are considered and discarded. One of these alternative models may actually be correct and is mistakenly discarded since it did not score as well as some other model. It would be useful to design an experiment that would provide data to either support the selected model or potentially an alternative model. As an example, consider a genetic circuit with three species and the three alternative models shown in Figure 2.20. If the selected model is the one shown in Figure 2.20(a) (i.e., no species influence the gene for C), then an experiment in which the gene that produces A is mutated to be non-functional could potentially be useful. This experiment provides more data where A is low, and if A actually does repress the gene for C as shown in Figure 2.20(b), this data would show more C expression. If A does not repress the gene for C, then the expression of C would not be affected by knocking out the gene for A. Another potentially useful experiment is one in which the gene that produces B is mutated to be non-functional. This experiment provides more data where B is low, and if both A and B repress the gene for C as shown in Figure 2.20(c) is correct, this data would show more C expression in those states in which both A and B are low. Finally, an experiment in which the genes that produces A and B are both mutated to be non-functional provides more data where both A and B are low. If it is indeed the case that both A and B can repress the gene for C, the resulting experimental data would show more C expression.

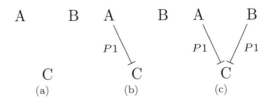

FIGURE 2.20:   Alternative models for experimental design example.
(a) No species influence expression level of the gene for C.
(b) Species A represses expression level of the gene C.
(c) Both species A and B repress expression level of the gene for C.

2.7   Sources

There are numerous excellent references that describe the experimental methods used today. This includes textbooks such as Berg *et al.* (2002); Watson *et al.* (2003); Alberts *et al.* (2002) as well as a variety of on-line resources such as `http://en.wikipedia.org`.

The idea of cluster analysis was proposed by the psychologist Robert Tryon in 1939 which he used to group individual people into clusters determined by their similarity in some measurable feature (Tryon, 1939). K-means clustering was proposed by MacQueen (1967). Hierarchical clustering was defined by Johnson (1967). Numerous researchers have applied clustering to gene expression data. For example, an application of K-means clustering to yeast gene expression data is described in Tavazoie *et al.* (1999). An application of agglomerative hierarchical clustering to gene expression data is described in Eisen *et al.* (1998). An excellent survey of clustering is given in D'haeseleer *et al.* (2000).

A useful tutorial on using Bayesian networks for learning is given in Heckerman (1996). Numerous researchers have been applying Bayesian networks to learning models of biological systems (Friedman *et al.*, 2000; Hartemink *et al.*, 2001; Pe'er, 2005; Sachs *et al.*, 2005). DBNs have also been successfully applied to numerous biological systems (Ong *et al.*, 2002; Husmeier, 2003; Nachman *et al.*, 2004; Yu *et al.*, 2004; Bernard and Hartemink, 2005; Beal *et al.*, 2005). Finally, the causal network learning method described in this chapter is an extension of Yu *et al.*'s DBN method that is due to Barker *et al.* which was originally described in (Barker *et al.*, 2006; Barker, 2007).

Problems

2.1 Generate synthetic time series data for the phage λ decision network model from Problem 1.8 using a stochastic simulator such as the one in `iBioSim`. Collect data for one run over five time points (5, 10, 15, 20, and 25 minutes). Use a clustering program such as `Cluster` and `Java TreeView` to cluster the data.

2.2 Generate synthetic time series data for the genetic circuit constructed in Problem 1.9 using a stochastic simulator such as the one in `iBioSim`. Collect data for one run over five time points (5, 10, 15, 20, and 25 minutes). Use a clustering program such as `Cluster` and `Java TreeView` to cluster the data.

2.3 The following questions use the data in this table:

lacI	*cI*	*tetR*	*gfp*	Probability
0	0	0	0	0.04
0	0	0	1	0.13
0	0	1	0	0.04
0	0	1	1	0.08
0	1	0	0	0.02
0	1	0	1	0.02
0	1	1	0	0.18
0	1	1	1	0.01
1	0	0	0	0.08
1	0	0	1	0.16
1	0	1	0	0.03
1	0	1	1	0.03
1	1	0	0	0.09
1	1	0	1	0.03
1	1	1	0	0.05
1	1	1	1	0.01

2.3.1. Based on the data above, select a Bayesian network.

2.3.2. Annotate each node with its corresponding conditional distribution.

2.3.3. Provide the equation for $P(lacI, cI, tetR, gfp)$ implied by the network.

2.3.4. Calculate the probability of each state using your Bayesian network. How well does this compare with the original probabilities?

2.3.5. Using this information, guess the causal network.

2.4   The following questions use the data in this table:

lacI	cI	tetR	gfp	Probability
0	0	0	0	0.08
0	0	0	1	0.03
0	0	1	0	0.03
0	0	1	1	0.06
0	1	0	0	0.10
0	1	0	1	0.05
0	1	1	0	0.14
0	1	1	1	0.06
1	0	0	0	0.04
1	0	0	1	0.11
1	0	1	0	0.03
1	0	1	1	0.08
1	1	0	0	0.06
1	1	0	1	0.06
1	1	1	0	0.04
1	1	1	1	0.03

2.4.1. Based on the data above, select a Bayesian network.

2.4.2. Annotate each node with its corresponding conditional distribution.

2.4.3. Provide the equation for $P(lacI, cI, tetR, gfp)$ implied by the network.

2.4.4. Calculate the probability of each state using your Bayesian network. How well does this compare with the original probabilities?

2.4.5. Using this information, guess the causal network.

2.5   Generate synthetic time series data for the phage λ decision network model from Problem 1.8 using a stochastic simulator such as the one in iBioSim. Collect data for 20 runs over 22 equally spaced time points from 0 to 2100 seconds. Use a Bayesian analysis program to learn a genetic circuit model.

2.6   Generate synthetic time series data for the genetic circuit constructed in Problem 1.9 using a stochastic simulator such as the one in iBioSim. Collect data for 20 runs over 22 equally spaced time points from 0 to 2100 seconds. Use a Bayesian analysis program to learn a genetic circuit model.

2.7   Generate synthetic time series data for the genetic circuit from Problem 1.3 using a stochastic simulator such as the one in iBioSim. Collect data for 20 runs over 22 equally spaced time points from 0 to 2100 seconds. Use the causal learning method in iBioSim to learn a genetic circuit model.

2.8   Generate synthetic time series data for the genetic circuit from Problem 1.4 using a stochastic simulator such as the one in iBioSim. Collect data for 20 runs over 22 equally spaced time points from 0 to 2100 seconds. Use the causal learning method in iBioSim to learn a genetic circuit model.

2.9   Generate synthetic time series data for the genetic circuit from Problem 1.5 using a stochastic simulator such as the one in iBioSim. Collect data for 20 runs over 22 equally spaced time points from 0 to 2100 seconds. Use the causal learning method in iBioSim to learn a genetic circuit model.

2.10   Generate synthetic time series data for the genetic circuit from Problem 1.6 using a stochastic simulator such as the one in iBioSim. Collect data for 20 runs over 22 equally spaced time points from 0 to 2100 seconds. Use the causal learning method in iBioSim to learn a genetic circuit model.

2.11   Generate synthetic time series data for the genetic circuit from Problem 1.7 using a stochastic simulator such as the one in iBioSim. Collect data for 20 runs over 22 equally spaced time points from 0 to 2100 seconds. Use the causal learning method in iBioSim to learn a genetic circuit model.

2.12   Generate synthetic time series data for the phage λ decision network model from Problem 1.8 using a stochastic simulator such as the one in iBioSim. Collect data for 20 runs over 22 equally spaced time points from 0 to 2100 seconds. Use the causal learning method in iBioSim to learn a genetic circuit model.

2.13   Generate synthetic time series data for the genetic circuit constructed in Problem 1.9 using a stochastic simulator such as the one in iBioSim. Collect data for 20 runs over 22 equally spaced time points from 0 to 2100 seconds. Use the causal learning method in iBioSim to learn a genetic circuit model.

Chapter 3

Differential Equation Analysis

Yes, we have to divide up our time like that, between our politics and our equations. But to me our equations are far more important, for politics are only a matter of present concern. A mathematical equation stands forever.

—Albert Einstein

After constructing a model for the biological system being studied, the next step of the engineering approach is to analyze this model. The goal of analysis is to be able to both reproduce experimental results *in silico* (i.e., in a computer simulation) as well as make predictions about the system for situations that have not yet been studied in the laboratory. Simulation has the benefit of providing unlimited controllability and observability allowing the computational systems biologist to potentially gain insight about the biological system being studied which would be difficult to achieve in a laboratory setting.

The traditional *classical chemical kinetics* (CCK) model utilizes *ordinary differential equations* (ODE) to represent system dynamics. As described in Chapter 1, the law of mass action can be used to translate a chemical reaction model into an ODE model. The differential equations derived in this way are known as *reaction rate equations*. Solving a set of ODEs is often extremely difficult, so one usually determines the behavior that they specify using numerical simulation. Such simulations, however, are complicated by the trade-off between simulation accuracy and efficiency. Such simulations also only yield results for one set of initial conditions and parameter values. Given the uncertainty in these values in biological systems, it is often quite useful to be able to perform qualitative analysis of an ODE model to better understand the system's behavior as these values vary. Finally, ODE models usually neglect spatial considerations. *Partial differential equations* (PDE) are typically utilized when spatial considerations are important.

This chapter is organized as follows. First, Section 3.1 describes the classical chemical kinetics ODE model. Next, Section 3.2 describes the basic concepts and methods for numerical simulation of such ODE models. Then, Section 3.3 presents methods for qualitative analysis of ODE models. Finally, Section 3.4 briefly presents modeling methods that incorporate spatial considerations.

3.1 A Classical Chemical Kinetic Model

A CCK model of a biochemical system tracks the concentrations of each chemical species (i.e., number of molecules divided by the volume, Ω, of the cell or compartment). A CCK model assumes that reactions occur in a *well-stirred* volume (i.e., molecules are equally distributed within the cell) which means that *spatial effects* (i.e., the location of the molecules within the cell) can be neglected. Finally, a CCK model assumes that reactions occur continuously and deterministically. For this last assumption to be acceptable, the number of molecules of each species must be large. Using these assumptions and the law of mass action, an ODE model can be derived to describe the dynamics of the biochemical system.

A CCK model is composed of n chemical species $\{S_1, \ldots, S_n\}$ and m chemical reaction channels $\{R_1, \ldots, R_m\}$. Each reaction R_j has the following form:

$$v_{1j}^r S_1 + \ldots + v_{nj}^r S_n \overset{k_f}{\underset{k_r}{\rightleftarrows}} v_{1j}^p S_1 + \ldots + v_{nj}^p S_n$$

where v_{ij}^r is the stoichiometry for species S_i as a reactant in reaction R_j and v_{ij}^p is the stoichiometry for species S_i as a product in reaction R_j. The values of v_{ij}^r and/or v_{ij}^p are 0 when species S_i does not participate as a reactant and/or product in reaction R_j. The parameter k_f is the *forward rate constant* while k_r is the *reverse rate constant*. If the reaction is *irreversible*, then k_r is 0.

The law of mass action states that the rate, or velocity, of an irreversible reaction is proportional to the product of the concentrations of the reactant molecules. The rate of a reversible reaction is also reduced by a value proportional to the product of the concentrations of product molecules. Formally, the reaction rate V_j for reaction R_j is defined as follows:

$$V_j = k_f \prod_{i=1}^{n} [S_i]^{v_{ij}^r} - k_r \prod_{i=1}^{n} [S_i]^{v_{ij}^p} \tag{3.1}$$

where $[S_i]$ is the concentration of species S_i. Using Equation 3.1, an ODE model can be constructed as follows:

$$\frac{d[S_i]}{dt} = \sum_{j=1}^{m} v_{ij} V_j, \quad 1 \leq i \leq n \tag{3.2}$$

where $v_{ij} = v_{ij}^p - v_{ij}^r$ (i.e., the net change in species S_i due to reaction R_j). In other words, the ODE model consists of one differential equation for each species which is the sum of the rates of change of the species due to each reaction that affects the species. As an example, the ODE model for the CI/CII portion of the phage λ decision circuit shown in Figure 3.1(a) is given in Figure 3.1(b).

$$
\begin{aligned}
\mathrm{CI} &\xrightarrow{k_d} () \\
\mathrm{CII} &\xrightarrow{k_d} () \\
P_{RE} + \mathrm{RNAP} &\xleftrightarrow{K_{o1}} S1 \\
P_R + \mathrm{RNAP} &\xleftrightarrow{K_{o2}} S2 \\
S1 &\xrightarrow{k_b} S1 + np\ \mathrm{CI} \\
S2 &\xrightarrow{k_o} S2 + np\ \mathrm{CII} \\
2\mathrm{CI} &\xleftrightarrow{K_d} \mathrm{CI}_2 \\
P_R + nc\ \mathrm{CI}_2 &\xleftrightarrow{K_r} S3 \\
P_{RE} + na\ \mathrm{CII} + \mathrm{RNAP} &\xleftrightarrow{K_a} S4 \\
S4 &\xrightarrow{k_a} S4 + np\ \mathrm{CI}
\end{aligned}
$$

Constant	*Value*
RNAP_0	$30\ nM$
K_d	$0.1M^{-1}$
K_{o1}	$0.01\ M^{-1}$
K_{o2}	$0.69422\ M^{-1}$
K_r	$0.2165\ M^{-nc}$
K_a	$0.00161\ M^{-(na+1)}$
k_o	$0.014\ \mathrm{sec}^{-1}$
k_b	$0.00004\ \mathrm{sec}^{-1}$
k_a	$0.015\ \mathrm{sec}^{-1}$
k_d	$0.0075\ \mathrm{sec}^{-1}$
nc	1
na	1
np	10

(a)

$$
\begin{aligned}
\frac{d[CI]}{dt} &= np\ k_b[S1] + np\ k_a[S4] - 2(K_d[CI]^2 - [CI_2]) - k_d[CI] \\
\frac{d[CI_2]}{dt} &= K_d[CI]^2 - [CI_2] - nc(K_r[P_R][CI_2]^{nc} - [S3]) \\
\frac{d[CII]}{dt} &= np\ k_o[S2] - na(K_a[P_{RE}][RNAP][CII]^{na} - [S4]) - k_d[CII] \\
\frac{d[P_R]}{dt} &= [S2] - K_{o2}[P_R][RNAP] + [S3] - K_r[P_R][CI_2]^{nc} \\
\frac{d[P_{RE}]}{dt} &= [S1] - K_{o1}[P_{RE}][RNAP] + [S4] - K_a[P_{RE}][RNAP][CII]^{na} \\
\frac{d[RNAP]}{dt} &= [S1] - K_{o1}[P_{RE}][RNAP] + [S2] - K_{o2}[P_R][RNAP] + \\
&\quad [S4] - K_a[P_{RE}][RNAP][CII]^{na} \\
\frac{d[S1]}{dt} &= K_{o1}[P_{RE}][RNAP] - [S1] \\
\frac{d[S2]}{dt} &= K_{o2}[P_R][RNAP] - [S2] \\
\frac{d[S3]}{dt} &= K_r[P_R][CI_2]^{nc} - [S3] \\
\frac{d[S4]}{dt} &= K_a[P_{RE}][RNAP][CII]^{na} - [S4]
\end{aligned}
$$

(b)

FIGURE 3.1: ODE model example. (a) Chemical reaction network model and (b) ODE model for the CI/CII portion of the phage λ decision circuit. Note that all reverse rate constants are assumed to be 1.0.

3.2   Differential Equation Simulation

A CCK model is a set of ODEs, or rate equations, in the following form:

$$\frac{dx_i}{dt} = f_i(\mathbf{x}), \text{ where } 1 \leq i \leq n \qquad (3.3)$$

where $\mathbf{x} = [x_1, \ldots, x_n] \geq \mathbf{0}$ is a vector of species concentrations. Solving such a set of differential equations analytically is typically very difficult, if not impossible. Therefore, numerical simulation is usually used instead. The goal of simulation is to approximate the function for the time evolution of the species concentrations, $\mathbf{X}(t)$, for $t \geq t_0$ assuming that the initial species concentrations $\mathbf{X}(t_0) = \mathbf{x}_0$. This is known as the *initial value problem*.

Many numerical methods have been proposed to solve the initial value problem. The simplest one is due to Euler. *Euler's method* firsts determines the initial instantaneous rate of change for each species, S_i, at time t_0 as follows:

$$\frac{dX_i(t_0)}{dt} = f_i(\mathbf{x}_0), \text{ where } 1 \leq i \leq n$$

If the rate of change returned by $f_i(\mathbf{X}(t))$ remains constant for all times $t \geq t_0$, then $X_i(t)$ equals $\mathbf{x}_{0i} + f_i(\mathbf{x}_0)(t - t_0)$. Unless f_i is a constant function, then $f_i(\mathbf{X}(t))$ does not remain constant for all time t, but it may be reasonable to assume that its value remains close to $f_i(\mathbf{x}_0)$ for some small time step Δt (known as the *step size*). With this assumption, simulation can determine the concentration of each species, S_i, at a time $t_1 = t_0 + \Delta t$ as follows:

$$X_i(t_1) \approx X_i(t_0) + f_i(\mathbf{X}(t_0))\Delta t$$

This equation can be generalized to compute the value of $X_i(t_j)$ for any $t_j = t_0 + n\Delta t$ where $j = 0, 1, 2, 3, \ldots$ as follows:

$$X_i(t_{j+1}) \approx X_i(t_j) + f_i(\mathbf{X}(t_j))\Delta t$$

Euler's method is illustrated with a simple example in Figure 3.2(a). At time t_0, the rate of change of species x_i is determined, and its value is changed at this rate until time t_1. Then, at time t_1, a new rate of change is calculated, and x_i is changed at this new rate until time t_2. This process continues until a simulation time limit is reached.

Since in Euler's method a new rate of change is calculated at each time point using only the current state information, this method is known as the *forward Euler method*, and it is an example of an *explicit* ODE simulation method. The *backward Euler method* is an example of an *implicit* ODE simulation method in that the new rate of change cannot be determined directly from the current state information. Instead, the new rate of change is given only

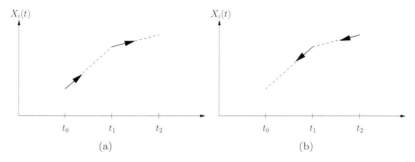

FIGURE 3.2: (a) An example illustrating the forward Euler method. (b) An example illustrating the backward Euler method.

implicitly by an equation that must be solved. In particular, the backward Euler method is defined by the following equation:

$$X_i(t_{j+1}) \approx X_i(t_j) + f_i(\mathbf{X}(t_{j+1}))\Delta t$$

and it is depicted in Figure 3.2(b). In other words, the new value of $\mathbf{X}(t_{j+1})$ is determined to be the one in which the rate at that point would have taken you there in a Δt step. Since the value $\mathbf{X}(t_{j+1})$ is not yet known, this equation must be solved using a root finding technique such as the Newton-Raphson method. While this makes an implicit method clearly much more complicated than an explicit method, implicit methods are often more stable for *stiff equations* (i.e., those that require a very small time step). This fact means that a similar quality result can be achieved using implicit methods with a much larger time step than would be required for explicit methods.

Using the fundamental theorem of calculus and Equation 3.3, the exact solution for each species, S_i, must satisfy:

$$X_i(t_{j+1}) = X_i(t_j) + \int_{t_j}^{t_{j+1}} f_i(\mathbf{X}(t))dt$$

The drawback of the Euler methods is that they approximate this integral by assuming that $f_i(\mathbf{X}(t))$ is constant throughout the entire Δt interval from t_j to t_{j+1}. Euler's method can be improved by approximating the rate of change in the Δt interval by the rate at the midpoint of that interval using the equation below:

$$X_i(t_{j+1}) \approx X_i(t_j) + \left[f_i\left(\mathbf{X}(t_j) + \frac{1}{2}f(\mathbf{X}(t_j))\Delta t\right) \right] \Delta t$$

This approach is known as the *midpoint method* or *second-order Runge-Kutta method*. The midpoint method is illustrated using Figure 3.3. In this method, the value of $\mathbf{X}$ is found halfway between t_0 and t_1 using the rates determined

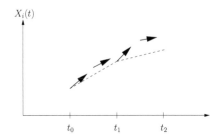

FIGURE 3.3: An example illustrating the midpoint, or second-order Runge-Kutta, method.

at t_0. Using this midpoint value, the rates are recalculated and this rate is used as the estimated rate from t_0 to t_1.

The midpoint method can be further improved by combining more points. For example, the *fourth-order Runge-Kutta method* is defined as follows:

$$\alpha_1 = f(\mathbf{X}(t_j))$$

$$\alpha_2 = f\left(\mathbf{X}(t_j) + \frac{\Delta t}{2}\alpha_1\right)$$

$$\alpha_3 = f\left(\mathbf{X}(t_j) + \frac{\Delta t}{2}\alpha_2\right)$$

$$\alpha_4 = f\left(\mathbf{X}(t_j) + \Delta t\alpha_3\right)$$

$$\mathbf{X}(t_{j+1}) = \mathbf{X}(t_j) + \frac{\Delta t}{6}\left[\alpha_1 + 2\alpha_2 + 2\alpha_3 + \alpha_4\right]$$

Implicit Runge-Kutta methods can also be used for stiff equations.

All the methods just described assume that the step size, Δt, is a constant, but in order to obtain good results, it is typically necessary to provide *adaptive stepsize control*. In other words, the simulation should slow down when rates are changing rapidly, and it should speed up when the rates are changing slowly. One approach is to use a technique known as *step doubling*. In this approach, the next state, $\mathbf{X}(t_{j+1})$, is calculated in one Δt step as well as another estimate of the next state, $\mathbf{X}'(t_{j+1})$, is determined by taking two $\Delta t/2$ steps. The difference between these two values represents an estimate of the error, E_i, for each species as defined below:

$$E_i = |X_i'(t_{j+1}) - X_i(t_{j+1})|$$

The goal of adaptive stepsize control is to adjust the stepsize to achieve a desired error level, D_i, for each species as defined below:

$$D_i = abs + rel \cdot |X_i(t_{j+1})|$$

where *abs* is the desired *absolute error level* and *rel* is the desired *relative error level*. If for any species, the error estimate, E_i, exceeds the desired error level, D_i, by more than 10 percent, the simulation step should be performed again using a new stepsize determined by the following equation:

$$\Delta t = 0.9 \cdot \Delta t \cdot \left(\frac{D}{E}\right)^q$$

where D/E is the minimum of the ratios D_i/E_i and q is the order of the method (i.e., 4 for a fourth-order Runge-Kutta method). If for all species, the error estimate is less than 50 percent of the desired error level, the stepsize can be increased using the following equation:

$$\Delta t = 0.9 \cdot \Delta t \cdot \left(\frac{D}{E}\right)^{(q+1)}$$

The previous simulation step can be accepted and the next simulation step uses this new stepsize. In this case, the state found by taking half steps, $\mathbf{X}'(t_{j+1})$, can be used as the new state as it is presumably more accurate.

A numerical simulation of the ODE model from Figure 3.1(b) for the CI/CII portion of the phage λ decision circuit is shown in Figure 3.4. Note that CI_total is the total number of CI molecules in either monomer or dimer form (i.e., CI_total $= 2 \cdot$ CI2 $+$ CI).

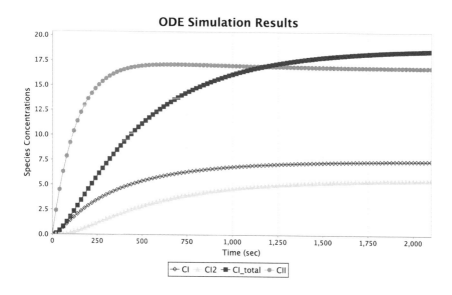

FIGURE 3.4: A numerical simulation of the ODE model from Figure 3.1(b) for the CI/CII portion of the phage λ decision circuit.

3.3   Qualitative ODE Analysis

The goal of ODE analysis is to determine properties of the *phase space* of the system being analyzed. The phase space is the set of all possible *states* of the system. The states in an ODE model are the values and current rates of change of each variable in the model. Numerical simulation of ODEs only show one trajectory in the phase space that can be reached from a given initial condition. Numerical simulation also considers only specific parameter values and potentially small changes in them may lead to substantial changes in qualitative behavior. The goal of *qualitative ODE analysis* is to determine the complete phase space based upon any initial condition. Qualitative ODE analysis is also employed to discover how parameter variation affects the phase space. It is quite an involved subject, so only a brief introduction is given here.

Let us first consider a *one-dimensional ODE model* with a single parameter, r, which has the following form:

$$\frac{dx}{dt} = f(x, r)$$

A state for such a model includes only the value of x and its current rate of change (or *flow*) of x, $\frac{dx}{dt}$. Therefore, a simple way to visualize its phase space is with a graph with x and $\frac{dx}{dt}$ as the two axes.

As an example, consider the following differential equation:

$$\frac{dx}{dt} = -r + x^2$$

The phase space for this example has three distinct forms depending on the value of r. When $r < 0$, the flow of x is always positive which means that x always increases without bound as shown in Figure 3.5(a). Note that the arrows on the x axis indicate the direction of x change in each distinct region. When $r = 0$, the flow of x is positive except when x is precisely 0 as shown in Figure 3.5(b). Such a point where there is no flow is called an *equilibrium (fixed) point*. An equilibrium point can be either *stable* or *unstable*. The *stability* of an equilibrium point is determined by considering the direction of flow near the point. In this case, the equilibrium point at $x = 0$ is unstable as any positive deviation away from this point would cause x to continue to increase moving it farther away from the equilibrium point. Finally, when $r > 0$, the flow of x is positive when $|x| > r$, negative when $|x| < r$, and 0 when $|x| = r$ as shown in Figure 3.5(c). In this case, there are two equilibrium points at $x = -r$ and $x = r$. The point at $x = -r$ is stable since any deviation away from this point drives it back to this point while the point at $x = r$ is unstable. In this example, as the parameter r is changed, the qualitative dynamics change substantially. The parameter value where this substantial change takes place ($r = 0$ in this example) is known as a *bifurcation point*.

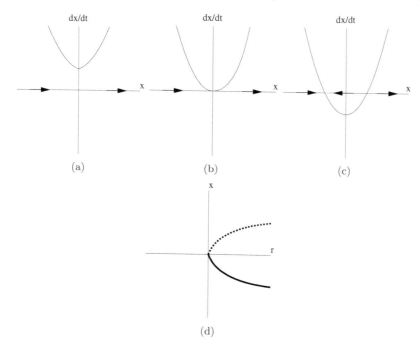

FIGURE 3.5: Saddle-node bifurcation example. (a) Phase space when $r < 0$. (b) Phase space when $r = 0$. (c) Phase space when $r > 0$. (d) Bifurcation diagram for this example.

A small change in the parameter value at this point results in a qualitative change in the phase space (i.e., a change in stability). In order to illustrate this affect of parameter variation on behavior, one can use a *bifurcation diagram* such as the one shown in Figure 3.5(d). A bifurcation diagram graphs the equilibrium points as a function of a parameter value. A solid line indicates a stable equilibrium point while a dashed line indicates an unstable one. This particular example illustrates a behavior known as a *saddle-node bifurcation*.

As another example, consider the following differential equation:

$$\frac{dx}{dt} = -rx + x^2$$

Again, in this example, there are three distinct forms depending on the value of r. When $r < 0$, as shown in Figure 3.6(a), there are two equilibrium points, a stable one at $x = -r$ and an unstable one at $x = 0$. When $r = 0$, there is only one unstable equilibrium point at $x = 0$ as shown in Figure 3.6(b). Finally, when $r > 0$, there are again two equilibrium points as shown in Figure 3.6(c), but this time the point at $x = 0$ is stable while the point at $x = r$ is unstable. These results are summarized in the bifurcation diagram in

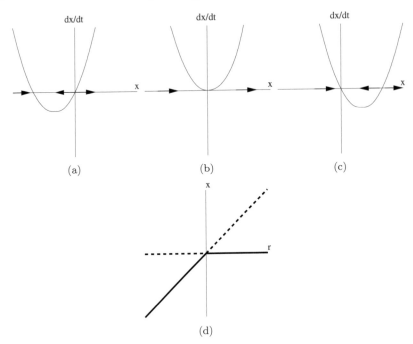

FIGURE 3.6: Transcritical bifurcation example. (a) Phase space when $r < 0$. (b) Phase space when $r = 0$. (c) Phase space when $r > 0$. (d) Bifurcation diagram for this example.

Figure 3.6(d). The point at $r = 0$ is again a bifurcation point for this example. This example illustrates a behavior known as a *transcritical bifurcation*.

As a final one-dimensional example, consider the following differential equation:

$$\frac{dx}{dt} = rx - x^3$$

In this example, when $r \leq 0$, the flow of x is positive when $x < 0$ and negative when $x > 0$ as shown in Figures 3.7(a) and 3.7(b). This result means there is one stable equilibrium point at $x = 0$. When $r > 0$, the equilibrium point at $x = 0$ becomes unstable, but two new stable equilibrium points appear at $x = -\sqrt{r}$ and $x = \sqrt{r}$ as shown in Figure 3.7(c). Again, $x = 0$ is a bifurcation point. This behavior known as a *pitchfork bifurcation* is illustrated in the bifurcation diagram shown in Figure 3.7(d).

A *two-dimensional ODE model* has the following form:

$$\frac{dx}{dt} = f(x, y)$$

$$\frac{dy}{dt} = g(x, y)$$

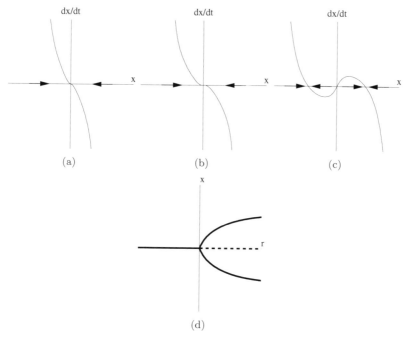

FIGURE 3.7: Pitchfork bifurcation example. (a) Phase space when $r < 0$. (b) Phase space when $r = 0$. (c) Phase space when $r > 0$. (d) Bifurcation diagram for this example.

In order to perform qualitative ODE analysis of two-dimensional systems, a different approach is needed to represent the phase space, since there are now four state variables (i.e., the values of x, y, and their flows). The graphical technique employed is to use *vector fields*. In particular, the graph uses the values of x and y as the axes, but it assigns a vector to each point indicating the direction of flow. This graph is constructed by first determining the *nullclines*. A nullcline is a line in which the flow of a variable is zero, and they are specified by the following algebraic equations:

$$0 = f(x, y)$$
$$0 = g(x, y)$$

To complete the graph, vectors are drawn at various points to indicate the direction of flow. The points in which the two nullclines intersect are the equilibrium points for the two-dimensional system. Again, their stability can be determined by looking at the flows represented by the vectors in the vicinity of the equilibrium point. Qualitative ODE analysis for systems of three or more dimensions becomes much more difficult due to the difficulty to visualize their phase space graphically. In these cases, one often applies assumptions to reduce the dimensionality of the system being analyzed until only two dimensions remain.

As an example, consider the following two-dimensional system:

$$\frac{d[CI]}{dt} = \frac{np\ P_{RE}RNAP(k_bK_{o1} + k_aK_a[CII])}{1 + K_{o1}RNAP + K_aRNAP[CII]} - k_d[CI]$$

$$\frac{d[CII]}{dt} = \frac{np\ k_oP_RK_{o2}RNAP}{1 + K_{o2}RNAP + K_rK_d[CI]^2} - k_d[CII]$$

This example is a reduction of the ODE model from Figure 3.1(b) to two dimensions that is found by assuming that the rates of change for the other dimensions are approximately zero. The nullcline for this model is shown in Figure 3.8. The dotted line in this plot represents the values of [CI] and [CII] in which the d[CI]/dt is equal to zero. Similarly, the solid line indicates where d[CII]/dt is zero. The point at which they intersect is an equilibrium point which matches well the results from Figure 3.4 (i.e., [CII] of about 18 and [CI] of about 8). The arrows represent the direction of flow at various points in the phase space with the length of the arrow indicating the magnitude of the flow. These arrows indicate that flow is into the equilibrium point, and it is smaller the closer it gets to the equilibrium point.

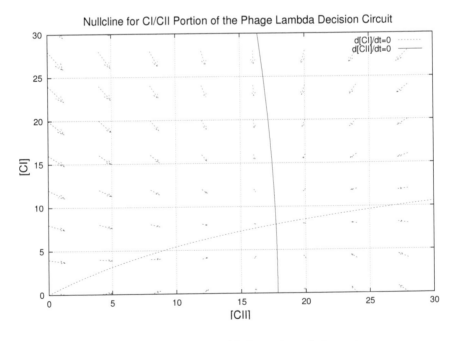

FIGURE 3.8: Nullcline for the CI/CII portion of phage λ.

3.4 Spatial Methods

As mentioned earlier, ODE models assume spatial homogeneity, but time delays due to diffusion or multiple compartments may be important. For example, diffusion between the nucleus and cytoplasm or other compartments may be important to model. Also, cell differentiation and embryonic development in multicellular organisms appears to be often controlled by gradients of protein concentrations across the cells.

One method of modeling these affects is to divide the system into regions as shown in Figure 3.9 where species are assigned to a region and movement between the regions is added to the ODE model. In particular, consider the configuration shown in Figure 3.9(a) in which p cells (or regions of a single cell) are arranged in a row in which each cell l has an amount of each species denoted by the vector $\mathbf{X}^l(t)$. Assuming that diffusion between cells is at a rate proportional to their differences in concentration (i.e., $x_i^{(l+1)} - x_i^{(l)}$ and $x_i^{(l-1)} - x_i^{(l)}$), the following *reaction-diffusion equations* can be obtained:

$$\frac{dx_i^{(l)}}{dt} = f_i(\mathbf{x}^{(l)}) + \delta_i(x_i^{(l+1)} - 2x_i^{(l)} + x_i^{(l-1)}), 1 \le i \le n, 1 < l < p \quad (3.4)$$

where δ_i is a diffusion constant. These equations are still ODEs, so they can be numerically solved using the simulation methods described earlier. These equations can also be generalized to more dimensions such as two-dimensions as shown in Figure 3.9(b).

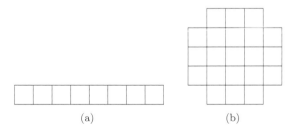

(a)   (b)

FIGURE 3.9: Examples of spatial configurations. (a) One-dimensional linear arrangement. (b) Two-dimensional configuration.

As the number of cells becomes large, l can be taken to be a continuous variable resulting in a partial differential equation (PDE) model:

$$\frac{dx_i}{dt} = f_i(\mathbf{x}) + \delta_i \frac{\partial^2 x_i}{\partial l^2}, 1 \le i \le n, 0 \le l \le \gamma \quad (3.5)$$

where the system size is assumed to be γ and diffusion does not occur beyond the boundaries at $l = 0$ and $l = \gamma$. However, analysis and numerical solutions for PDE models is substantially more involved. Interested readers are referred to the references given below for applications of PDEs to the modeling of biological systems.

3.5   Sources

In 1864, Waage and Guldberg introduced the law of mass action which is the basis of classical chemical kinetics (Waage and Guldberg (1864), translated in Abrash (1986)). There are whole books dedicated to classical chemical kinetics such as Wright (2004). Chemical kinetic models of genetic circuits were first proposed in Goodwin (1963, 1965). An excellent reference on numerical simulation of ODE models can be found in Press *et al.* (1992). Qualitative ODE analysis is also the subject of many textbooks such as Strogatz (1994).

An ODE model of phage λ can be found in Reinitz and Vaisnys (1990). There have also been ODE models developed for various other networks such as the *lac* operon (Wong *et al.*, 1997; Vilar *et al.*, 2003; Yildirim and Mackey, 2003; Yildirim *et al.*, 2004; Santillán and Mackey, 2004; Hoek and Hogeweg, 2006), bacteriophage T7 (Endy *et al.*, 2000; You and Yin, 2006), *trpyptophan* synthesis in *E. coli* (Giona and Adrover, 2002; Xiu *et al.*, 2002; Santillán and Zeron, 2004, 2006), circadian rhythms (Leloup and Goldbeter, 1951; Ruoff *et al.*, 2001; Ueda *et al.*, 2001), and the cell cycle (Novak and Tyson, 1995; Tyson *et al.*, 1996; Borisuk and Tyson, 1998; Novak *et al.*, 1998; Tyson, 1999; Chen *et al.*, 2000).

Reaction-diffusion equations for modeling cell differentiation were originally proposed by Turing (1951). The goal of his work is a mathematical model of a growing embryo. Turing suggested that systems consist of masses of tissues within which certain substances called *morphogens* react chemically and diffuse. Their diffusing into a tissue persuades the tissue to develop differently than it would have without it being present. The embryo in the spherical blastula stage has spherical symmetry. Systems with spherical symmetry whose state is changed by chemical reactions and diffusion remains spherically symmetric. This cannot result in a non-spherically symmetric organism like a horse. There are, however, *some* asymmetries which cause instability and lead to a new and stable equilibrium without symmetry. This behavior is very similar to how electrical oscillators get started. The most successful applications of reaction-diffusion equations and other spatial models have been to modeling pattern formation in the *Drosophila* embryo (see for example Kauffman *et al.* (1978); Bunow *et al.* (1980); Goodwin and Kauffman (1990); Lacalli (1990); Myasnikova *et al.* (2001)).

Problems

3.1 Consider the following reactions:

$$2S_1 \xrightarrow{0.1} 2S_2$$
$$S_1 + S_2 \xrightarrow{0.2} 2S_1$$

3.1.1. Determine the reaction rate equations for $[S_1]$ and $[S_2]$.

3.1.2. Simulate by hand using Euler's method the following set of differential equations for 1 second with a time step of 0.2 seconds starting with initial concentrations of $[S_1] = 3.0$ and $[S_2] = 5.0$.

3.1.3. Redo your simulation using the Fourth-Order Runge-Kutta.

3.2 Consider the following reactions:

$$S_1 + S_2 \xrightarrow{0.1} S_3$$
$$S_3 \xrightarrow{0.2} S_1 + S_4$$
$$S_4 \xrightarrow{0.3} S_2$$

3.2.1. Determine the reaction rate equations for $[S_1]$, $[S_2]$, $[S_3]$, and $[S_4]$.

3.2.2. Simulate by hand using Euler's method the following set of differential equations for 1 second with a time step of 0.2 seconds starting with initial concentrations of $[S_1] = 10.0$, $[S_2] = 10.0$, $[S_3] = 10.0$, and $[S_4] = 10.0$.

3.2.3. Redo your simulation using the Fourth-Order Runge-Kutta.

3.3 Consider the following reactions:

$$2A \underset{k_2}{\overset{k_1}{\rightleftharpoons}} A_2$$

$$O + A_2 + \text{RNAP} \underset{k_4}{\overset{k_3}{\rightleftharpoons}} C \xrightarrow{k_5} O + A_2 + \text{RNAP} + P$$

3.3.1. Determine the reaction rate equations for $[A]$, $[A_2]$, $[O]$, $[\text{RNAP}]$, $[C]$, and $[P]$.

3.3.2. Simulate by hand using Euler's method your reaction rate equations for 0.1 second with a time step of 0.01 seconds starting with initial concentrations of $(O, RNAP, A, A_2, C, P) = (1, 30, 10, 0, 0, 0)$ and assuming $k_1 = 1$, $k_2 = 0.1$, $k_3 = 1$, $k_4 = 1$, and $k_5 = 0.1$.

3.4 Derive an ODE model for the reaction-based model from Problem 1.3.

3.5 Derive an ODE model for the reaction-based model from Problem 1.4.

3.6 Derive an ODE model for the reaction-based model from Problem 1.5.

3.7 Derive an ODE model for the reaction-based model from Problem 1.6.

3.8 Using `iBioSim` or another tool that supports differential equation simulation, simulate the reaction-based model from Problem 1.3 with a Runge-Kutta method (ex. rkf45) with a time limit of 2100 and print interval of 50. Make a note of the simulation time and plot CI_2, LacI, and TetR. Next, simulate using Euler's method. Repeat the Euler simulation using different time steps until the results match up well. What time step is required for a good match? How do the simulation times compare?

3.9 Using `iBioSim` or another tool that supports differential equation simulation, simulate the reaction-based model from Problem 1.4 with a Runge-Kutta method (ex. rkf45) with a time limit of 2100 and print interval of 50. Make a note of the simulation time and plot CI_2 and Cro_2. Next, simulate using Euler's method. Repeat the Euler simulation using different time steps until the results match up well. What time step is required for a good match? How do the simulation times compare?

3.10 Using `iBioSim` or another tool that supports differential equation simulation, simulate the reaction-based model from Problem 1.5 with a Runge-Kutta method (ex. rkf45) with a time limit of 2100 and print interval of 50. Make a note of the simulation time and plot CI_2, LacI, and TetR. Next, simulate using Euler's method. Repeat the Euler simulation using different time steps until the results match up well. What time step is required for a good match? How do the simulation times compare?

3.11 Using `iBioSim` or another tool that supports differential equation simulation, simulate the reaction-based model from Problem 1.6 with a Runge-Kutta method (ex. rkf45) with a time limit of 2100 and print interval of 50. Make a note of the simulation time and plot CI_2, LacI, TetR, and GFP. Next, simulate using Euler's method. Repeat the Euler simulation using different time steps until the results match up well. What time step is required for a good match? How do the simulation times compare?

3.12 Using `iBioSim` or another tool that supports differential equation simulation, simulate the CI/CII portion of the phage λ model from Problem 1.7 with a Runge-Kutta method (ex. rkf45) with a time limit of 2100 and print interval of 50. Make a note of the simulation time and plot CI_2 and CII. Next, simulate using Euler's method. Repeat the Euler simulation using different time steps until the results match up well. What time step is required for a good match? How do the simulation times compare?

3.13 Using `iBioSim` or another tool that supports differential equation simulation, simulate the phage λ decision circuit model from Problem 1.8 with a Runge-Kutta method (ex. rkf45) with a time limit of 2100 and print interval of 50. Make a note of the simulation time and plot CI_2 and Cro_2. Next, simulate using Euler's method. Repeat the Euler simulation using different time steps until the results match up well. What time step is required for a good match? How do the simulation times compare?

3.14 Using iBioSim or another tool that supports differential equation simulation, simulate the genetic circuit model from Problem 1.9 with a Runge-Kutta method (ex. rkf45) with a time limit of 2100 and print interval of 50. Make a note of the simulation time and plot the important species. Next, simulate using Euler's method. Repeat the Euler simulation using different time steps until the results match up well. What time step is required for a good match? How do the simulation times compare?

3.15 Implement Euler's method using your favorite programming language.

3.16 Implement a Fourth-order Runge-Kutta method using your favorite programming language.

3.17 Create a nullcline plot for LacI and TetR using the equations below:

$$\frac{d[\text{LacI}]}{dt} = \frac{10k_oO_tK_o\text{RNAP}_0}{1 + K_o\text{RNAP}_0 + K_r[\text{TetR}]^2} - k_d[\text{LacI}]$$

$$\frac{d[\text{TetR}]}{dt} = \frac{10k_oO_tK_o\text{RNAP}_0}{1 + K_o\text{RNAP}_0 + K_r[\text{LacI}]^2} - k_d[\text{TetR}]$$

where $k_o = 0.05$, $O_t = 1$, $K_o = 0.033$, $\text{RNAP}_0 = 30$, $K_r = 0.25$, and $k_d = 0.0075$. This plot should include three equilibrium points. Identify the stable and unstable points. The equations above use the same rates for both LacI and TetR. Try using different values for these parameters in the two equations, and identify values where the number of equilibrium points is different.

Chapter 4

Stochastic Analysis

That which is static and repetitive is boring. That which is dynamic and random is confusing. In between lies art.

—John A. Locke

A philosopher once said "It is necessary for the very existence of science that the same conditions always produce the same results." Well, they do not. You set up the circumstances, with the same conditions every time, and you cannot predict behind which hole you will see the electron.

—Richard Feynman

As described in Chapter 3, a chemical reaction model can be transformed into a set of first order ODEs. An ODE model, however, assumes that concentrations vary continuously and deterministically. Unfortunately, chemical systems satisfy neither of these assumptions. The number of molecules of a chemical species is a discrete quantity in that it is always a natural number (0, 1, 2, ...). Furthermore, chemical reactions typically occur after two molecules collide. Unless one tracks the exact position and velocity of every molecule (something one obviously does not wish to do), it is impossible to know when a reaction may occur. Therefore, it is preferable to consider the occurrence of chemical reactions to be a stochastic process. Despite the violation of these assumptions, in systems which involve large molecular counts, ODE models give a quite accurate picture of their behavior. If the molecular counts are small, however, the discrete and stochastic nature may have significant influence on the observable behavior. Genetic circuits typically involve small molecule counts, since there are often only a few 10s or 100s of molecules of each transcription factor and one strand of DNA. Therefore, accurate analysis of genetic circuits often requires a stochastic process description.

This chapter presents one such description. First, Section 4.1 describes the *stochastic chemical kinetic* (SCK) model. Next, Section 4.2 describes the *chemical master equation* (CME), the formal representation that describes the time evolution of the probabilities for the states of an SCK model. Sections 4.3 to 4.5 present algorithms to numerically analyze the CME. Section 4.6 describes the relationship between the CME and the reaction rate equations. Section 4.7 introduces *stochastic Petri nets* (SPN), an alternative modeling formalism for stochastic models. Section 4.8 presents a stochastic model of the phage λ decision circuit. Finally, Section 4.9 briefly describes a method to consider spatial issues in a stochastic model.

4.1   A Stochastic Chemical Kinetic Model

Recall that a chemical reaction network model is composed of n chemical species $\{S_1, \ldots, S_n\}$ and m chemical reaction channels $\{R_1, \ldots, R_m\}$. An SCK model assumes that the species are contained within some constant volume Ω such as within a cell. An SCK model also assumes that the system is *well-stirred*. A well-stirred system is one in which the molecules are equally distributed within the cell. This assumption allows one to neglect spatial effects (i.e., the location of the molecules within the cell). This chapter presents methods that account for spatial effects later. An SCK model also assumes that the system is in *thermal equilibrium* (i.e., it maintains a constant temperature, T), but not necessarily *chemical equilibrium*.

If $X_i(t)$ is the number of molecules of S_i at time t, then the vector $\mathbf{X}(t) = (X_1(t), \ldots, X_n(t))$ is the state of the system at time t. $\mathbf{X}(t_0) = \mathbf{x}_0$ is the initial number of molecules for some initial time t_0. If a reaction, R_μ, occurs, the state, $\mathbf{x}$, is updated by adding to it the *state-change vector*, $\mathbf{v}_\mu$ (i.e., $\mathbf{x}' = \mathbf{x} + \mathbf{v}_\mu$). The state-change vector, $\mathbf{v}_\mu$, is equal to $(v_{1\mu}, \ldots, v_{n\mu})$ where $v_{i\mu}$ is the change in the molecule count of S_i due to the reaction R_μ. The two-dimensional array $\{v_{i\mu}\}$ is also known as the *stoichiometric matrix*. A reaction, R_μ, is said to be *elemental* if the reaction can be considered a distinct physical event that happens nearly instantaneously. For elemental reactions, the values of $v_{i\mu}$ are practically constrained to the values of $0, \pm 1, \pm 2$.

Every reaction channel R_μ has associated with it a *specific probability rate constant*, c_μ, which is related as described later to the reaction rate constant, k_μ. The value of c_μ is selected such that $c_\mu dt$ can be defined to be the probability that a randomly chosen combination of reactant molecules react as defined by the R_μ reaction inside the volume Ω in the next infinitesimal time interval $[t, t+dt]$. Multiplying c_μ by the number of possible combinations of reactant molecules for R_μ in a state $\mathbf{x}$ yields the *propensity function*, a_μ. In other words, $a_\mu(\mathbf{x})dt$ is defined to be the probability that R_μ occurs in the state $\mathbf{x}$ within Ω in the next infinitesimal time interval $[t, t+dt]$.

As an example, consider a *bimolecular reaction channel*, R_μ, which has the following form:

$$S_1 + S_2 \xrightarrow{c_\mu} S_3 + \ldots$$

To determine c_μ for this reaction, it is necessary to find the probability that a S_1 molecule and S_2 molecule collide and react within the next dt time units. This derivation assumes that each S_i is a hard sphere of mass m_i with radius r_i. The assumption of thermal equilibrium means that a randomly selected molecule of S_i can be found uniformly distributed within the volume Ω. It also means that the average relative speed in which S_2 sees S_1 moving is given by the formula $\bar{v}_{12} = \sqrt{8 k_B T / \pi m_{12}}$ where k_B is Boltzmann's constant and $m_{12} = m_1 m_2 / (m_1 + m_2)$. In the next dt time interval, the S_2 molecule sweeps a collision cylinder relative to S_1 which has a height $\bar{v}_{12}dt$ and base area $\pi(r_1+r_2)^2$. Since the locations of the molecules are uniformly distributed within Ω, the probability that S_1 is within the collision cylinder is the ratio of $\pi(r_1+r_2)^2\bar{v}_{21}dt$ to the volume Ω. Therefore, the specific probability rate constant is given by the formula:

$$c_\mu = \Omega^{-1}\pi(r_1+r_2)^2\bar{v}_{21}p_\mu$$

where p_μ is the probability that S_1 and S_2 react when they collide. Assuming that S_1 and S_2 react only when their kinetic energy exceeds the activation energy, ϵ_μ, then c_μ is given by this formula:

$$c_\mu = \Omega^{-1}\pi(r_1+r_2)^2\left(\frac{8k_BT}{\pi m_{12}}\right)^{1/2}exp(-\epsilon_\mu/k_BT) \qquad (4.1)$$

The number of possible combinations of S_1 and S_2 molecules that can react is simply x_1x_2, so the propensity function for R_μ is $a_\mu(\mathbf{x}) = c_\mu x_1 x_2$. If, however, the bimolecular reaction has two identical molecules as reactants (i.e., $S_1 + S_1 \rightarrow S_2 \dots$), then the number of distinct combinations of S_1 molecules is $x_1(x_1 - 1)/2$. The propensity function in this case is $a_\mu(\mathbf{x}) = c_\mu x_1(x_1 - 1)/2$.

A *monomolecular reaction channel* has the form:

$$S_1 \overset{c_\mu}{\rightarrow} S_2 + \dots$$

In such a reaction, the reactant molecule, S_1, makes a spontaneous change in its internal structure. To determine c_μ for this type of reaction requires quantum mechanical considerations. The propensity function for this reaction is simply $a_\mu(\mathbf{x}) = c_\mu x_1$. If this reaction is actually an enzymatic reaction of the form:

$$E + S_1 \overset{c_\mu}{\rightarrow} E + S_2 + \dots$$

where E is an enzyme, this reaction should be considered as a bimolecular reaction.

A *trimolecular reaction channel* has the form:

$$S_1 + S_2 + S_3 \overset{c_\mu}{\rightarrow} S_4 + \dots$$

Such a reaction, however, is typically used only as an approximation because the probability of three molecules colliding within an infinitesimal time interval dt is extremely small. Therefore, a trimolecular (or greater) reaction channel is usually an approximation for a sequence of reactions such as those below:

$$S_1 + S_2 \underset{c_2}{\overset{c_1}{\rightleftharpoons}} S^* \text{ and } S^* + S_3 \overset{c_3}{\rightarrow} S_4 + \dots$$

Such an approximation may be reasonable when the lifetime of S^* is very short (i.e., $1/c_2$ is very small). In this case, the probability that a molecule of

S^* during its short lifetime reacts with a randomly chosen molecule of S_3 is approximately $c_3(1/c_2)$. Next, let us consider a small but finite time interval Δt which is still much larger than the lifetime of S^* (i.e., $\Delta t \gg 1/c_2$). If Δt is sufficiently small, then the probability that S_1 and S_2 react in that time interval to form S^* is $c_1\Delta t$. Combining the probabilities of the first and third reaction determines that the probability of both reactions occurring in the time interval Δt would be $(c_1c_3/c_2)\Delta t$. In other words, c_μ for the trimolecular reaction approximation of these two channels can be taken to be:

$$c_\mu = c_1c_3/c_2 \tag{4.2}$$

This result, however, is an approximation because Δt is not a true infinitesimal. The propensity function for this reaction is $a_\mu(\mathbf{x}) = c_\mu x_1 x_2 x_3$.

As shown in Equation 4.1, c_μ is proportional to Ω^{-1} for bimolecular reactions. This makes sense because the larger the volume, the less likely that two molecules collide. A monomolecular reaction, however, is independent of volume since these reactions are performed on a single reactant spontaneously. Finally, as shown in Equation 4.2, it is proportional to Ω^{-2} since it is the product of two bimolecular reaction constants and divided by a monomolecular reaction constant. In general,

$$c_\mu \propto \Omega^{-(m-1)}$$

where m is the number of reactant molecules in R_μ. This relationship is also key in understanding the relationship between the specific probability rate constant, c_μ, and the reaction rate constant, k_μ. For monomolecular reactions, c_μ is equal to k_μ. For bimolecular reactions, c_μ is equal to k_μ/Ω if the reactants are different species and $2k_\mu/\Omega$ if the same species.

4.2 The Chemical Master Equation

The stochastic model just described is a *jump Markov process*. A *Markov process* is one where the next state is only dependent on the present state and not the past history. A jump Markov process is one in which the state updates occur in discrete amounts. Due to the stochastic nature of the model, it is not possible to know the exact state $\mathbf{X}(t)$, but rather only the probability of being in a given state at time t starting from a state $\mathbf{X}(t_0) = \mathbf{x}_0$ (i.e., $\mathcal{P}(\mathbf{x}, t|\mathbf{x}_0, t_0)$). Using probability theory, an equation can be devised to describe this probability using a time-evolution of step dt as shown below:

$$\mathcal{P}(\mathbf{x}, t + dt | \mathbf{x}_0, t_0) = \mathcal{P}(\mathbf{x}, t | \mathbf{x}_0, t_0) \times \left[1 - \sum_{j=1}^{m} (a_j(\mathbf{x}) dt) \right]$$

$$+ \sum_{j=1}^{m} \mathcal{P}(\mathbf{x} - \mathbf{v}_j, t | \mathbf{x}_0, t_0) \times (a_j(\mathbf{x} - \mathbf{v}_j) dt)$$

The first half of the sum on the right is the probability that it is already in the state $\mathbf{x}$ at time t, and there is no reaction in the time period $[t, t + dt]$. The second half indicates that it is $\mathbf{v}_j$ away from state $\mathbf{x}$ at time t, and the reaction R_j occurs in the time period $[t, t + dt]$. Note that dt is chosen to be small enough that at most one reaction can occur during this time period. Performing the limit shown below results in the CME which defines the time evolution of state probabilities, $\mathcal{P}(\mathbf{x}, t | \mathbf{x}_0, t_0)$.

$$\frac{\partial \mathcal{P}(\mathbf{x}, t | \mathbf{x}_0, t_0)}{\partial t} = \lim_{dt \to 0} \frac{\mathcal{P}(\mathbf{x}, t + dt | \mathbf{x}_0, t_0) - \mathcal{P}(\mathbf{x}, t | \mathbf{x}_0, t_0)}{dt}$$

$$= \sum_{j=1}^{m} [a_j(\mathbf{x} - \mathbf{v}_j) \mathcal{P}(\mathbf{x} - \mathbf{v}_j, t | \mathbf{x}_0, t_0) - a_j(\mathbf{x}) \mathcal{P}(\mathbf{x}, t | \mathbf{x}_0, t_0)]$$

Unfortunately, this differential equation cannot be solved analytically or numerically except in the very simplest situations since it represents a set of equations that is nearly as large as the number of molecules in the system.

4.3 Gillespie's Stochastic Simulation Algorithm

Although the CME usually cannot be solved analytically or numerically, trajectories for $\mathbf{X}(t)$ can be generated using *stochastic simulation*. One approach would be to follow the approach taken in differential equation simulation which is to pick a small time step, dt, and at each time step update the system state by either selecting a reaction to occur or doing nothing. One readily observes, however, that for a sufficiently small dt, the vast majority of time steps result in no reaction.

Gillespie's stochastic simulation algorithm (SSA) improves the efficiency of simulation by stepping over all of these useless time steps. This algorithm is not based directly on the CME, but rather an equivalent formulation that uses a new probability function $p(\tau, \mu | \mathbf{x}, t)$. This function is defined such that $p(\tau, \mu | \mathbf{x}, t) d\tau$ is the probability that the next reaction in the system is R_μ which occurs in the infinitesimal time interval $[t + \tau, t + \tau + d\tau]$ assuming that the current state is $\mathbf{X}(t) = \mathbf{x}$. Therefore, this function is a joint probability density function for two random variables, τ (time to the next reaction) and μ (the index of the next reaction) given the fact that the system is in state $\mathbf{x}$ at time t. The advantage of this formulation is that the simulation advances from one reaction to the next skipping over any time points in which no reaction occurs.

To construct an algorithm that uses this function, let us first derive an analytical expression for $p(\tau, \mu | \mathbf{x}, t)$. First, let us introduce a new function, $\mathcal{P}_0(\tau | \mathbf{x}, t)$, that represents the probability that there is no reaction in the time interval $[t, t + \tau)$. Using this function, the function $p(\tau, \mu | \mathbf{x}, t)$ can be defined as follows:

$$p(\tau, \mu | \mathbf{x}, t)d\tau = \mathcal{P}_0(\tau | \mathbf{x}, t) \times (a_\mu(\mathbf{x})d\tau) \tag{4.3}$$

In other words, no reactions occur in the interval $[t, t + \tau)$ and the R_μ reaction occurs in the interval $[t + \tau, t + \tau + d\tau]$. The function $\mathcal{P}_0(\tau | \mathbf{x}, t)$ must satisfy the following:

$$\mathcal{P}_0(\tau + d\tau | \mathbf{x}, t) = \mathcal{P}_0(\tau | \mathbf{x}, t) \times \left[1 - \sum_{j=1}^{m} (a_j(\mathbf{x})d\tau) \right]$$

Using this formula, the following differential equation can be derived:

$$\frac{d\mathcal{P}_0(\tau, \mathbf{x}, t)}{d\tau} = -a_0(\mathbf{x})\mathcal{P}_0(\tau | \mathbf{x}, t) \text{ where } a_0(\mathbf{x}) = \sum_{j=1}^{m} a_j(\mathbf{x})$$

This differential equation with the initial condition $\mathcal{P}_0(\tau = 0 | \mathbf{x}, t) = 1$ has the following solution:

$$\mathcal{P}_0(\tau | \mathbf{x}, t) = exp(-a_0(\mathbf{x})\tau) \tag{4.4}$$

Inserting Equation 4.4 into Equation 4.3 and canceling $d\tau$ yields:

$$p(\tau, \mu | \mathbf{x}, t) = exp(-a_0(\mathbf{x})\tau) \times a_\mu(\mathbf{x}) \tag{4.5}$$

which can be rewritten as:

$$p(\tau, \mu | \mathbf{x}, t) = a_0(\mathbf{x})exp(-a_0(\mathbf{x})\tau) \times \frac{a_\mu(\mathbf{x})}{a_0(\mathbf{x})} \tag{4.6}$$

This form shows that $p(\tau, \mu | \mathbf{x}, t)$ can be divided into a probability density function for τ and another for μ. The variable τ is an exponential random variable with mean and standard deviation of $1/a_0(\mathbf{x})$. The variable μ is an integer random variable with point probabilities $a_\mu(\mathbf{x})/a_0(\mathbf{x})$.

Using these observations, Gillespie devised the SSA algorithm shown in Figure 4.1. In step 1, this algorithm sets the current state of the system to be the starting time t_0 and the starting molecule counts $\mathbf{x}_0$. At the beginning of each iteration (step 2 of the SSA algorithm), all the propensity functions, $a_j(\mathbf{x})$ and their sum must be recalculated, since state changes caused by the last reaction may have caused these propensities to change. Of course, not all propensities change or change significantly during a single iteration. Using this observation, several modifications to this algorithm described later have been proposed that improve the algorithm's efficiency significantly. In step 3, two uniform random numbers, r_1 and r_2, are selected between $[0, 1]$. These random numbers are used in the next two steps to determine τ and μ. In step 4, τ is found using the inversion generating method which makes uses of the following relationship:

$$\int_0^\tau a_0(\mathbf{x})exp(-a_0(\mathbf{x})\tau')d\tau' = 1 - r_1$$

In step 5, μ is found by summing the propensities until the sum exceeds $r_2 a_0(\mathbf{x})$. In step 6, the new state is found by moving time forward τ time units and updating the molecule counts using the stoichiometry information for the reaction, $\mathbf{v}_\mu$. If the current time, t, is larger than the desired simulation time, then in step 7 this trajectory is complete and simulation halts. Otherwise, in step 8, the new state is recorded and simulation continues back at step 2.

1. Initialize: $t = t_0$ and $\mathbf{x} = \mathbf{x}_0$.

2. Evaluate propensity functions $a_j(\mathbf{x})$ at state $\mathbf{x}$, and also their sum $a_0(\mathbf{x}) = \sum_{j=1}^m a_j(\mathbf{x})$.

3. Draw two unit uniform random numbers, r_1 and r_2.

4. Determine the time, τ, until the next reaction:

$$\tau = \frac{1}{a_0(\mathbf{x})} \ln\left(\frac{1}{r_1}\right)$$

5. Determine the next reaction, R_μ:

$$\mu = \text{ the smallest integer satisfying } \sum_{j=1}^\mu a_j(\mathbf{x}) > r_2 a_0(\mathbf{x})$$

6. Determine the new state after reaction μ: $t = t + \tau$ and $\mathbf{x} = \mathbf{x} + \mathbf{v}_\mu$.

7. If t is greater than the desired simulation time then halt.

8. Record $(\mathbf{x}, t)$ and go to step 2.

FIGURE 4.1: Gillespie's stochastic simulation algorithm (SSA).

Using SSA to compute a single trajectory is no more complex to implement than numerical simulation used to solve reaction rate equations. It also provides a closer approximation of molecular reality for systems with small molecule counts such as genetic circuits. As an example, Figure 4.2 shows a comparison of SSA and ODE simulation for the total amount of CI (i.e., CI_total) from the CI/CII portion of the phage λ decision circuit. In these simulations, the initial number of molecules for P_R and P_{RE} are set to 1 (i.e., a single phage DNA) and RNAP is 30. This figure shows that the SSA and ODE results actually vary significantly. In fact, CI_total actually oscillates in the SSA simulation while it reaches an equilibrium in the ODE simulation.

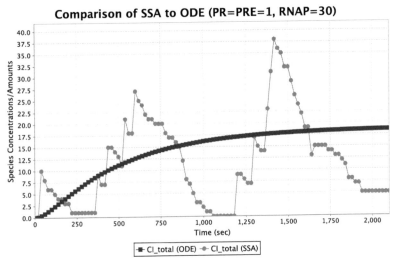

FIGURE 4.2: Comparison of SSA to ODE simulation for CI_total from the CI/CII portion of the phage λ decision circuit. Initial number of molecules for $P_R = 1$, $P_{RE} = 1$, and RNAP= 30.

Unfortunately, the SSA has a substantial computational cost for two reasons. First, SSA must be run many times (often 1000s) to generate many trajectories to produce reasonable statistics on the system's behavior. Numerical simulations of reaction rate equations are deterministic, so they only need to be run once. Second, SSA is often very slow since the mean time step, τ, is equal to $1/a_0(\mathbf{x})$ and $a_0(\mathbf{x})$ can be very large when any molecule counts become large. In this case, reactions are so closely spaced that it requires a prohibitively large amount of computational effort. When molecule counts increase, however, the relative difference between deterministic and stochastic trajectories decrease. Figure 4.3 shows the result of SSA and ODE simulation with 10 P_R and P_{RE} molecules, and 300 RNAP molecules. This figure shows that with these larger molecule counts that the SSA and ODE simulations are in much better agreement.

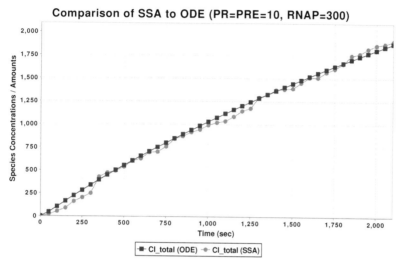

FIGURE 4.3: Comparison of SSA to ODE simulation for CI_total from the CI/CII portion of the phage λ decision circuit. Initial number of molecules for $P_R = 10$, $P_{RE} = 10$, and RNAP= 300.

4.4 Gibson/Bruck's Next Reaction Method

The SSA algorithm described in the last section in which the next reaction and its time are calculated explicitly is known as the *direct method*. Gillespie also proposed another method known as the *first reaction method* shown in Figure 4.4. In this approach, a next reaction time, τ_j, is calculated for every reaction, R_j. The reaction, R_μ, with the smallest next reaction time, τ_μ, is found, and this *first reaction* is executed.

On the surface, the first reaction method appears to be a very inefficient approach as it requires the generation of m random variables per simulation step as opposed to the two generated by the SSA algorithm. This algorithm, however, can be improved using the observation made earlier that not all propensities change as the result of a reaction. Gibson and Bruck used this

1. Initialize: $t = t_0$ and $\mathbf{x} = \mathbf{x}_0$.

2. Evaluate propensity functions $a_j(\mathbf{x})$ at state $\mathbf{x}$.

3. For each j, determine the time, τ_j, until the next R_j reaction:

$$\tau_j = \frac{1}{a_j(\mathbf{x})} \ln\left(\frac{1}{r_j}\right).$$

where each r_j is a unit uniform random number.

4. Let R_μ be the reaction whose τ_μ is the smallest.

5. Let τ equal τ_μ.

6. Determine the new state after reaction R_μ: $t = t + \tau$ and $\mathbf{x} = \mathbf{x} + \mathbf{v}_\mu$.

7. If t is greater than the desired simulation time then halt.

8. Record $(\mathbf{x}, t)$ and go to step 2.

FIGURE 4.4: Gillespie's first reaction method.

observation to devise a new method based on the first reaction method which they call the *next reaction method*. In particular, the following three steps are taken during every iteration and take a time proportional to the number of reactions, m.

1. Update all m propensity functions, $a_j(\mathbf{x})$.

2. Generate m random numbers, r_j, and next reaction times, τ_j.

3. Find the smallest reaction time, τ_μ.

The next reaction method eliminates each of these performance bottlenecks by using clever data structures to avoid recalculating propensities and next reaction times when this is unnecessary.

First, the τ_j and $a_j(\mathbf{x})$ values are stored, so they can be used in subsequent iterations. Second, the τ_j values are changed to use absolute time rather than relative time between reactions. This makes it possible for the τ_j values to be useful for more than one iteration. Third, a *dependency graph* is created that indicates which reactions affect which other reactions when they are executed. Fourth, this algorithm reuses every τ_j except the one for the reaction being executed, τ_μ. This reuse is accomplished by re-normalizing the τ_j value whenever its propensity has changed. Finally, an *indexed priority queue* is introduced to organize the $a_j(\mathbf{x})$ and τ_j data such that it is easy to update and easy to find the smallest entry. The complete next reaction method is shown in Figure 4.5.

1. Initialize:

   (a) $t = t_0$ and $\mathbf{x} = \mathbf{x}_0$.

   (b) Generate a dependency graph, G.

   (c) Evaluate propensity functions $a_j(\mathbf{x})$ at state $\mathbf{x}$.

   (d) For each j, determine the time, τ_j, until the next R_j reaction:

   $$\tau_j = t + \frac{1}{a_j(\mathbf{x})} \ln\left(\frac{1}{r_j}\right)$$

   where each r_j is a unit uniform random number.

   (e) Store the τ_j values in an indexed priority queue Q.

2. Let R_μ be the reaction whose τ_μ is the smallest stored in Q.

3. Let τ equal τ_μ.

4. Determine the new state after reaction R_μ: $t = \tau$ and $\mathbf{x} = \mathbf{x} + \mathbf{v}_\mu$.

5. For each edge (μ, α) in the dependency graph G,

   (a) Set $a_{\alpha,old} = a_\alpha$ and update a_α.

   (b) If $\alpha \neq \mu$, set $\tau_\alpha = (a_{\alpha,old}/a_\alpha)(\tau_\alpha - t) + t$.

   (c) If $\alpha = \mu$, generate a random number, r_μ, and

   $$\tau_\mu = t + \frac{1}{a_\mu(\mathbf{x})} \ln\left(\frac{1}{r_\mu}\right)$$

6. If t is greater than the desired simulation time then halt.

7. Record $(\mathbf{x}, t)$ and go to step 2.

FIGURE 4.5: Gibson and Bruck's next reaction method.

In a dependency graph, there is a vertex for each reaction, R_j, and an edge, (j, k), from this vertex to every other vertex, R_k, that has as a reactant either a reactant or product of R_j. The dependency graph for the CI/CII portion of the phage λ decision circuit is shown in Figure 4.6. To simplify presentation, the reversible reactions are shown as single reactions though they must be split into irreversible reactions before stochastic analysis. Also, double arrows are used to indicate that two reactions mutually depend upon each other, and self-loops are omitted since every reaction depends on itself. This graph indicates, among other things, that when the CI dimerization reaction, r_7, fires, the only propensities that must be updated, other than the one for r_7, are for the CI degradation reaction, r_1, and the CI_2 binding reaction for repression of P_R, r_8 (technically only the forward reaction for r_8).

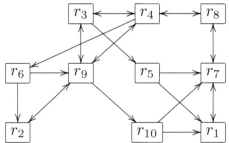

FIGURE 4.6:   Dependency graph for the CI/CII portion of the phage λ decision circuit.

An indexed priority queue is a tree structure in which the parent always has a lower τ_j value than both its children. This fact means the top node always has the smallest τ_j value. Care, of course, must be taken to preserve this property as the τ_j values are updated. To facilitate this update, there is also an index array that is used to locate a reaction within the priority queue. In particular, when a τ_j value is updated, the corresponding reaction, R_j is looked up in the index array to find its location in the priority queue. If after the update, the new value is smaller than its parent, this reaction must swap locations in the tree with its parent. If instead the new value is greater than its smallest child, this reaction must swap places in the tree with this child. If the reaction is moved to a new location, the update must continue recursively to see if the reaction must propagate further up or down the tree. This update function is quite efficient, since it can always be done in $O(log(m))$ time where m is the number of reactions. An example indexed priority queue for the CI/CII portion of the phage λ decision circuit is shown in Figure 4.7. Note that there are entries for both the forward reaction, r_{if}, and reverse reaction, r_{ir}, for reversible reactions. This queue indicates that the next reaction to fire should be the binding of RNAP to the promoter P_R (i.e., r_{4f}). Using the dependency graph in Figure 4.6, the algorithm determines that the reaction times for r_3, r_6, r_8 and r_9 must be updated in addition to finding a new reaction time for r_{4f}. After each reaction time is updated, its position in the tree may need to be updated. For example, assume that the new reaction time for r_6 is found to be 10.1. To keep the queue consistent, this would require reaction r_6 to swap places with reaction r_{3f} followed by swapping places with r_{7r}.

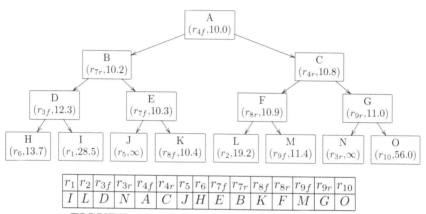

FIGURE 4.7: Indexed priority queue example.

4.5 Tau-Leaping

While the next reaction method does improve the efficiency significantly for systems with many species and many reaction channels, every reaction event must still be simulated one at a time. For many interesting systems, this is simply not practical. For such systems, it is necessary to give up some of the exactness to improve the simulation speed. One such method is known as *tau-leaping*. In the SSA algorithm, the value of τ is selected such that it jumps precisely to the next reaction. In tau-leaping, this restriction is lifted meaning that many reactions are allowed to fire at once in the time interval $[t, t + \tau]$. This change is accomplished by introducing m random functions, $K_j(\tau, \mathbf{x}, t)$, where each one returns the number of times that the reaction channel, R_j, fires in the time interval $[t, t + \tau]$ assuming the system is currently in the state $\mathbf{X}(t) = \mathbf{x}$. Using these functions, the new state after this τ-leap is determined by the formula below:

$$\mathbf{X}(t + \tau) = \mathbf{x} + \sum_{j=1}^{m} K_j(\tau, \mathbf{x}, t)\mathbf{v}_j$$

Unfortunately, these functions are not easy to compute since they are all dependent on each other. Namely, the number of reactions of a reaction channel, R_j, is dependent on its propensity, $a_j(\mathbf{x})$, which is dependent on the state which is dependent on the number of all the other reactions. Even if the joint probability density function could be computed, it would likely be more expensive than doing the full simulation in the first place, defeating the purpose.

The approach taken, therefore, is to approximate the values for $K_j(\tau, \mathbf{x}, t)$ using an assumption known as the *leap condition*. The leap condition states that the value of τ must be chosen to be small enough such that no propensity function changes by a significant amount. If the leap condition is satisfied, then the value of $K_j(\tau, \mathbf{x}, t)$ for each reaction, R_j, can be approximated to be a statistically independent *Poisson random variable*:

$$K_j(\tau, \mathbf{x}, t) \approx \mathcal{P}_j(a_j(\mathbf{x}), \tau) \quad (j = 1, \ldots, m)$$

where $\mathcal{P}_j(a_j(\mathbf{x}), \tau)$ returns the number of events k in the interval $[t, t+\tau]$ such that the probability of each k value is governed by the probability:

$$\mathcal{P}[k \text{ events}] = \frac{e^{-a_j(\mathbf{x})\tau}(a_j(\mathbf{x})\tau)^k}{k!}$$

The crucial question is how to find a τ that is small enough to satisfy the leap condition, but large enough to fire enough events to speedup simulation.

One method for selecting τ uses the equation below:

$$\tau = \min_{i \in I_{rs}} \left\{ \frac{max\{\epsilon_i x_i, 1\}}{|\sum_{j \in J_{ncr}} v_{ij}a_j(\mathbf{x})|}, \frac{max\{\epsilon_i x_i, 1\}^2}{\sum_{j \in J_{ncr}} v_{ij}^2 a_j(\mathbf{x})} \right\} \quad (4.7)$$

where I_{rs} are the species that appear as reactants in reactions and J_{ncr} are the *non-critical reactions*. A reaction is non-critical if it can be fired n_c times (a value typically between 5 and 30) without causing one of its reactant species counts to become negative. The goal of this equation is to ensure that no propensity function is likely to change by more than $\epsilon a_j(\mathbf{x})$, where ϵ is an *accuracy control parameter* satisfying $0 < \epsilon << 1$. The value of ϵ_i is ϵ for a species, S_i, that only appears as a reactant in unimolecular reactions, $\epsilon/2$ if it only appears in bimolecular reactions involving two different species, and $\epsilon/(2+(x_i-1)^{-1})$ if it appears in bimolecular reactions involving two molecules of the same species.

The tau-leaping algorithm is shown in Figure 4.8. The value of ϵ selected provides a means of trading off simulation accuracy for runtime. For a large

ϵ, a significant runtime improvement can be achieved at the cost of some accuracy. Care has to be taken though as large jumps can cause bad things such as species counts being made negative. The tau-leaping algorithm avoids this by identifying those critical reactions that are close to exhausting one of their reactants. At most one critical reaction is allowed to fire per iteration of the algorithm. As ϵ is made smaller, tau-leaping gradually reduces to the SSA. For a very small ϵ, however, it is not as efficient as the SSA as it begins to take many τ leaps that produce no events. In fact, if τ is found to be less than a few multiples of $1/a_0(\mathbf{x})$, then the algorithm should revert to the SSA to allow the algorithm to step to the next reaction.

1. Initialize: $t = t_0$ and $\mathbf{x} = \mathbf{x}_0$.

2. Evaluate propensity functions $a_j(\mathbf{x})$ at state $\mathbf{x}$.

3. Determine J_{ncr}.

4. If $J_{ncr} = \emptyset$ then $\tau' = \infty$ else determine value for τ' using Equation 4.7.

5. If J_{ncr} includes all reactions then $\tau'' = \infty$ else use SSA to compute τ'' and j_c, the next critical reaction.

6. $\tau = min(\tau', \tau'')$ and $t = t + \tau$.

7. $\mathbf{x} = \mathbf{x} + \sum_{j \in J_{ncr}} \mathcal{P}_j\left(a_j(\mathbf{x})\tau\right) \mathbf{v}_j$.

8. If $\tau'' \leq \tau'$ then $\mathbf{x} = \mathbf{x} + \mathbf{v}_{j_c}$.

9. If t is greater than the desired simulation time then halt.

10. Record $(\mathbf{x}, t)$ and go to step 2.

FIGURE 4.8: Explicit tau-leaping simulation algorithm.

4.6 Relationship to Reaction Rate Equations

The previous section stated that if time step $\Delta t = \tau$ is chosen to be small enough such that no propensity function changes significantly, then the state update can be approximated using m statistically independent Poisson random variables.

$$\mathbf{X}(t + \Delta t) \approx \mathbf{x} + \sum_{j=1}^{m} \mathcal{P}_j(a_j(\mathbf{x}), \Delta t)\mathbf{v}_j \tag{4.8}$$

Assuming that Δt is chosen to be large enough so that there are many firings of each reaction (i.e., $a_j(\mathbf{x})\Delta t >> 1$ for all $j = 1, \ldots, m$), the Poisson variable can be approximated with a normal random variable, $\mathcal{N}_j(\text{mean}, \text{variance})$, with the same mean and variance.

$$\mathbf{X}(t + \Delta t) \approx \mathbf{x} + \sum_{j=1}^{m} \mathcal{N}_j(a_j(\mathbf{x})\Delta t, a_j(\mathbf{x})\Delta t)\mathbf{v}_j$$

$$= \mathbf{x} + \sum_{j=1}^{m} \mathbf{v}_j a_j(\mathbf{x})\Delta t + \sum_{j=1}^{m} \mathbf{v}_j \sqrt{a_j(\mathbf{x})}\mathcal{N}_j(0, 1)\sqrt{\Delta t}$$

using the fact that $\mathcal{N}(m, \sigma^2) = m + \sigma\mathcal{N}(0, 1)$. Finally, assuming that Δt is a *macroscopically infinitesimal* time increment dt results in the following:

$$\mathbf{X}(t + dt) \approx \mathbf{X}(t) + \sum_{j=1}^{m} \mathbf{v}_j a_j(\mathbf{X}(t))dt + \sum_{j=1}^{m} \mathbf{v}_j \sqrt{a_j(\mathbf{X}(t))}N_j(t)\sqrt{dt} \tag{4.9}$$

where $N_j(t)$ are m statistically independent and temporally uncorrelated normal random variables with mean 0 and variance 1. Equation 4.9 is known as the *chemical Langevin equation*.

Equation 4.9 is composed of two parts: a deterministic component that grows linearly with respect to the propensity functions and a stochastic component that grows proportional to the square root of the propensity functions. The propensity functions grow in direct proportion with the the size of the system (species populations and volume) which means that the stochastic component scales relative to the deterministic component as the inverse square root of the system size. Therefore, as the size of a system increases, the magnitude of the stochastic fluctuations diminish. At some point, these fluctuations become insignificant meaning Equation 4.9 can be simplified to:

$$\mathbf{X}(t + dt) \approx \mathbf{X}(t) + \sum_{j=1}^{m} \mathbf{v}_j a_j(\mathbf{X}(t))dt \tag{4.10}$$

Rearranging Equation 4.10 results in the following:

$$\frac{\mathbf{X}(t + dt) - \mathbf{X}(t)}{dt} = \frac{d\mathbf{X}(t)}{dt} = \sum_{j=1}^{m} \mathbf{v}_j a_j(\mathbf{X}(t)) \tag{4.11}$$

This equation is simply the reaction rate equation, but it has been derived from stochastic chemical kinetics. Therefore, the reaction rate equation is valid (as shown in the simulation earlier) when the system is large enough such that no propensity function changes significantly in a dt time step and every reaction channel fires many times within that time step.

4.7 Stochastic Petri-Nets

One interesting graphical model that has been applied to represent biological systems are *Petri nets* (no relation to Petri dishes although their places look kind of like them). Petri nets were invented by Carl Adam Petri at the age of 13 ironically for the purpose of representing chemical processes, and they have since been used in a wide variety of applications including the modeling of circuits, software, communication networks, and manufacturing processes. Stochastic Petri nets (SPNs) are a variant of Petri nets that are isomorphic to jump Markov processes. Using SPNs as a modeling formalism has the advantage that several tools have been developed for their analysis.

An SPN is composed of a set of *places*, P, a set of *transitions*, T, an *input function*, I, an *output function*, O, a *weight function*, W, and an *initial marking*, M_0. An SPN representation for the reaction to form water is shown in Figure 4.9. The places are represented using circles, and they correspond to molecular species in a chemical reaction network. In Figure 4.9, there are three places to model the three species, H_2, O_2, and H_2O. The transitions, T, are represented using boxes, and they correspond to reactions. This simple SPN has one transition representing the one reaction. The input function, I, takes a pair (p, t) where p is a place and t is a transition, and it returns a natural number, v. If v is greater than 0, then an arc is drawn from p to t. In this case, the species associated with p is a reactant for the reaction associated with t, and v is its stoichiometry. If v is greater than 1, the arc is labeled with v. The input function for this simple SPN specifies that H_2 and O_2 are reactants with H_2 having a stoichiometry of 2. The output function takes a pair (t, p), and it returns a natural number, v. If v is greater than 0, then an arc is drawn from t to p, and it indicates that species p is a product of reaction t. The output function for this simple SPN specifies that H_2O is a product, and it has stoichiometry of 2. The weight function assigns a rate to a transition, and it is used here to specify the rate of a reaction. For simplicity, the transition is labeled with the rate constant, but it is assumed that the rate function includes the product of the reactant concentrations. For this simple SPN, the weight function would be $2k[H_2]^2[O_2]$ with the transition labeled with k. The state of an SPN is its *marking* which is an assignment of a number of tokens to each place in the net. The marking corresponds to the current number of each molecular species. The initial marking, M_0, represents the initial token counts and thus the initial numbers of molecules. The initial marking for this simple SPN indicates that there are initially four molecules of H_2 and three molecules of O_2.

It is not too difficult to translate a reaction-based model of any regulatory network into an SPN model. A portion of the reactions involved in the P_{RE} promoter for phage λ is shown in Figure 4.10 along with the reactions for dimerization and degradation of CI. The behavior specified with this SPN

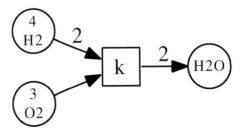

FIGURE 4.9:   A simple SPN for the formation of water.

would begin with the firing of the transition labeled Ka which would remove one token from each of the places for RNAP, CII, and PRE. At this point, there are two possibilities, the transition labeled 1 could fire returning the net to the initial state or the transition labeled ka could fire resulting in 10 tokens being placed in CI representing the production of 10 molecules of CI. Note that this reaction is connected to S4 with a double arrow which indicates that the transition replaces the token immediately after consuming it. In other words, S4 is a modifier for this reaction. In this state, there are now several possible next transitions, S4 could break back down into its parts, more CI could be produced, two molecules of CI could dimerize, or a molecule of CI could degrade. A tool which simulates SPNs would choose the transition using a stochastic simulation similar to one of the methods described earlier in this chapter where the likelihood of each transition would be governed by its rate.

4.8   Phage λ Decision Circuit Example

As described in Chapter 1, phage λ has two developmental pathways to replicate its DNA. The decision as to which pathway is to be followed appears to be stochastic. It is controlled by two independently produced regulatory proteins that are competing for a genetic switch. The result is that the switch behavior is stochastic in that within a homogeneous population of phage λ infected *E. coli* under the same environmental conditions some cells undergo lysis while others become a lysogen. A deterministic model always results in exactly one possible outcome unless the parameters or initial conditions of the model are changed. Therefore, stochastic analysis as described in this chapter is necessary in order to predict the probability that a cell heads down the lysis or lysogeny pathway after infection.

The goal of stochastic analysis is to predict the probability of lysogeny under various conditions. In particular, the probability of lysogeny has been shown experimentally to be dependent on the number of phage particles to

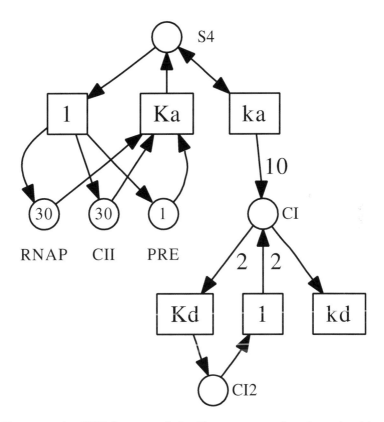

FIGURE 4.10:   An SPN for part of the P_{RE} promoter for phage λ with the reactions for dimerization and degradation of CI.

infect a cell, or multiplicity of infection (MOI), and on the nutritional state of the cell. Namely, the higher the MOI, the more likely the cell ends up in lysogeny. The reason for this is that the proteases in a cell that degrade proteins are constant in MOI while the number of genes producing CII and CIII, the proteins crucial to establishing lysogeny, are proportional to MOI. In other words, lysogeny which depends on several proteins building up to get established benefits from this net increase in production. It has also been shown that well-fed cells tend to go into lysis. The reason for this is that the proteases are more active in well-fed cells shortening the lifetimes of CII and CIII. The stochastic analysis of the model should reproduce these effects.

The decision between the lysis and lysogeny pathway is essentially determined by a race between the buildup of the proteins Cro_2 and CI_2. On the one hand, if Cro_2 reaches its critical level first, then the cell undergoes lysis. On the other hand, if CI_2 reaches its critical level first, then the cell becomes a lysogen. Figure 4.11(a) shows the time evolution of the amounts of the Cro_2 and CI_2 proteins for the average of 100 stochastic simulation runs for the phage λ decision circuit model from Chapter 1 assuming an MOI of 3. This plot seems to indicate that for an MOI of 3 the lysogeny pathway is always taken. However, Figures 4.11(b) and (c) show the result of two stochastic analysis runs on the same model. While in Figure 4.11(b) the lysogeny pathway is indeed chosen, the result shown in Figure 4.11(c) indicates that the lysis pathway has been chosen.

In 1973, Kourilsky demonstrated in the laboratory that the lysogenic fraction increased with *average phage input* (API). It is difficult in the lab to determine exactly how many viruses infect a given *E. Coli* cell (i.e., its MOI). Therefore, instead one measures the proportion of phage to bacteria. An API of 0.1 means there is one virus per ten *E. Coli* while an API of 10.0 means there are ten viruses to every one *E. Coli*. His experiments were performed on O^- and P^- strains of phage λ making them incapable of phage chromosome replication. This makes it unnecessary to include phage chromosome replication in the model. The experiments were performed on both starved and well-fed cells. Stochastic analysis is only practical for the starved data as the number of simulation runs goes like $1/f$ where f is the fraction of lysogens. Figure 4.12 shows the probability of lysogeny versus MOI, $F(M)$, found by simulating our model for 10,000 simulation runs using SSA.

From MOI data, API data can be obtained using a Poisson distribution:

$$\mathcal{P}(M, A) = \frac{A^M}{M!} e^{-A}$$

$$F_{\text{lysogen}}(A) = \sum_M \mathcal{P}(M, A) \cdot F(M)$$

The results shown in Figure 4.13 are found to match the experimental data quite well. More important than replicating experimental data, a model should also make predictions. Therefore, Figure 4.14 shows predictions made

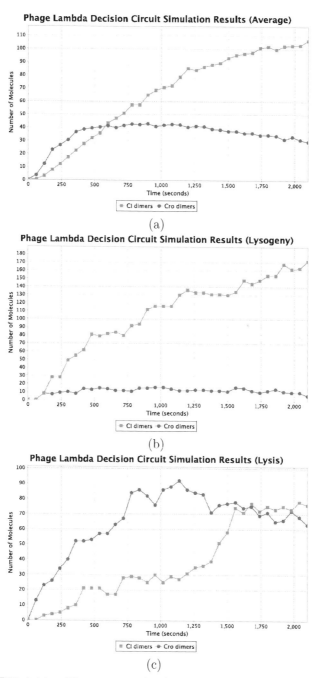

FIGURE 4.11: Time courses for CI_2 and Cro_2: (a) average (b) run resulting in lysogeny, and (c) run resulting in lysis.

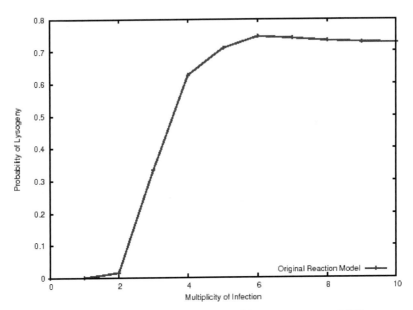

FIGURE 4.12:   Probability of lysogeny versus MOI
(courtesy of Kuwahara *et al.* (2006)).

by our model for *O-T-* mutants (phages without terminator switches) and
O-N- mutants (phages without the *N* gene). The results are as expected in
that phages without terminators take the lysogeny pathway more often while
phages without the *N* gene end up in lysis more often.

4.9   Spatial Gillespie

All the methods presented so far have the well-stirred assumption. As the
volume of the system being analyzed increases, this assumption becomes less
valid. There have been several methods proposed to add spatial considerations
to stochastic simulation. In the *spatial Gillespie method*, the system is divided
into several discrete subvolumes as shown in Figure 4.15. The size of the
subvolumes is selected such that within a subvolume that the well-stirred
assumption is reasonable. In other words, diffusion within the subvolume is
faster than the rate of the reactions in the model. The state of the system now
is the number of each species within each subvolume. During each simulation
cycle, a molecule can either react with other molecules within a subvolume,
or a molecule can diffuse to an adjacent subvolume. Another way of looking

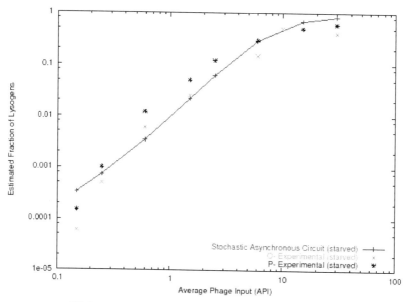

FIGURE 4.13: Lysogenic fraction versus API (courtesy of Kuwahara *et al.* (2006)).

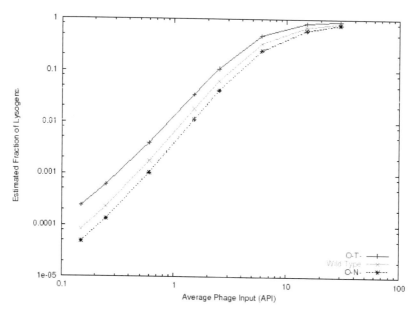

FIGURE 4.14: Lysogenic fraction versus API for *O-T-* and *O-N-* mutants (courtesy of Kuwahara *et al.* (2006)).

at this is if you begin with a system which has species, $(S_1, \ldots, S_n)$, spatial considerations can be added by dividing these species into $(S_1^{(i,j,k)}, \ldots, S_n^{(i,j,k)})$ where $i = 1, \ldots, p$, $j = 1, \ldots, q$, and $l = 1, \ldots, r$, assuming that the volume has been divided into $p \times q \times r$ subvolumes. Finally, to the set of reactions, these reactions would be added:

$$S_\mu^{(i,j,l)} \overset{k}{\leftrightarrow} S_\mu^{(i+1,j,l)}$$

$$S_\mu^{(i,j,l)} \overset{k}{\leftrightarrow} S_\mu^{(i,j+1,l)}$$

$$S_\mu^{(i,j,l)} \overset{k}{\leftrightarrow} S_\mu^{(i,j,l+1)}$$

After making these modifications to the model, then any of the stochastic simulation algorithms presented in this chapter can be used for analysis.

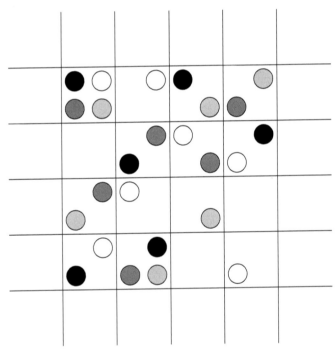

FIGURE 4.15: Discrete subvolumes used by the spatial Gillespie method.

4.10   Sources

In his seminal paper (Gillespie, 1977), Gillespie introduced the idea of using a stochastic process description for chemical reactions with small species counts and developed the SSA algorithm. Much of the material in the first six sections of this chapter are inspired by the descriptions given in Gillespie (1992, 2005). More details about Gibson and Bruck's improvements on the SSA algorithm can be found in Gibson and Bruck (2000). The tau-leaping algorithm is described in Gillespie and Petzold (2003); Cao *et al.* (2006), and an implicit version to overcome dynamically stiff systems is described in Rathinam *et al.* (2003). A stochastic simulation method for rare events is described in Kuwahara and Mura (2008).

SPNs were originally proposed in Molloy (1982), and they were later generalized to include immediate transitions that fire after no delay (Marsan *et al.*, 1984). A detailed discussion of SPNs can be found in Marsan *et al.* (1994). SPNs were applied to modeling biological systems in Goss and Peccoud (1998). The original stochastic analysis of the phage λ decision circuit that appeared in Arkin *et al.* (1998) was the first analysis of a complete system that demonstrated the importance of using stochastic analysis for genetic circuits. An excellent survey of various methods that can be used to extend stochastic modeling to include spatial considerations is given in Takahashi *et al.* (2005). The spatial Gillespie method presented in Section 4.9 is from Stundzia and Lumsden (1996).

Problems

4.1   Simulate by hand using Gillespie's SSA the following set of reactions assuming $(S_1, S_2) = (4, 4)$. Be sure to show all steps.

$$2S_1 \xrightarrow{0.1} 2S_2$$
$$S_1 + S_2 \xrightarrow{0.2} 2S_1$$

Create three separate trajectories with three reactions in each trajectory (if three are possible).

4.2   Simulate by hand using Gillespie's SSA the following set of reactions assuming $(S_1, S_2, S_3, S_4) = (10, 10, 10, 10)$. Be sure to show all steps.

$$S_1 + S_2 \xrightarrow{0.1} S_3$$
$$S_3 \xrightarrow{0.2} S_1 + S_4$$
$$S_4 \xrightarrow{0.3} S_2$$

Create three separate trajectories with three reactions in each trajectory.

4.3 In this problem, use a tool such as iBioSim that supports both ODE and SSA simulation. Simulate the reaction-based model from Problem 1.3 using ODE simulation with a time limit of 2100 and print interval of 50, and note the simulation time. Repeat using SSA simulation and a run count of 1. Again, note the simulation time. Plot the results for CI_2 for both the ODE and SSA simulations. Repeat the SSA simulation using a run count of 10, and plot the average for CI_2 with the ODE results for CI_2. Repeat again using a run count of 100. How do the simulation times compare as run counts increase? Do your ODE and SSA simulations agree as you increase the number of runs? Does ODE simulation appear to be reasonably accurate for this example?

4.4 In this problem, use a tool such as iBioSim that supports both ODE and SSA simulation. Simulate the reaction-based model from Problem 1.4 using ODE simulation with a time limit of 2100 and print interval of 50, and note the simulation time. Repeat using SSA simulation and a run count of 1. Again, note the simulation time. Plot the results for CI_2 for both the ODE and SSA simulations. Repeat the SSA simulation using a run count of 10, and plot the average for CI_2 with the ODE results for CI_2. Repeat again using a run count of 100. How do the simulation times compare as run counts increase? Do your ODE and SSA simulations agree as you increase the number of runs? Does ODE simulation appear to be reasonably accurate for this example?

4.5 In this problem, use a tool such as iBioSim that supports both ODE and SSA simulation. Simulate the reaction-based model from Problem 1.5 using ODE simulation with a time limit of 2100 and print interval of 50, and note the simulation time. Repeat using SSA simulation and a run count of 1. Again, note the simulation time. Plot the results for CI_2 for both the ODE and SSA simulations. Repeat the SSA simulation using a run count of 10, and plot the average for CI_2 with the ODE results for CI_2. Repeat again using a run count of 100. How do the simulation times compare as run counts increase? Do your ODE and SSA simulations agree as you increase the number of runs? Does ODE simulation appear to be reasonably accurate for this example?

4.6 In this problem, use a tool such as iBioSim that supports both ODE and SSA simulation. Simulate the reaction-based model from Problem 1.6 using ODE simulation with a time limit of 2100 and print interval of 50, and note the simulation time. Repeat using SSA simulation and a run count of 1. Again, note the simulation time. Plot the results for CI_2 for both the ODE and SSA simulations. Repeat the SSA simulation using a run count of 10, and plot the average for CI_2 with the ODE results for CI_2. Repeat again using a run count of 100. How do the simulation times compare as run counts increase? Do your ODE and SSA simulations agree as you increase the number of runs? Does ODE simulation appear to be reasonably accurate for this example?

4.7   In this problem, use a tool such as `iBioSim` that supports both ODE and SSA simulation. Simulate the reaction-based model from Problem 1.7 using ODE simulation with a time limit of 2100 and print interval of 50, and note the simulation time. Repeat using SSA simulation and a run count of 1. Again, note the simulation time. Plot the results for CI_2 for both the ODE and SSA simulations. Repeat the SSA simulation using a run count of 10, and plot the average for CI_2 with the ODE results for CI_2. Repeat again using a run count of 100. How do the simulation times compare as run counts increase? Do your ODE and SSA simulations agree as you increase the number of runs? Does ODE simulation appear to be reasonably accurate for this example?

4.8   In this problem, use a tool such as `iBioSim` that supports both ODE and SSA simulation. Simulate the reaction-based model from Problem 1.8 using ODE simulation with a time limit of 2100 and print interval of 50, and note the simulation time. Repeat using SSA simulation and a run count of 1. Again, note the simulation time. Plot the results for CI_2 for both the ODE and SSA simulations. Repeat the SSA simulation using a run count of 10, and plot the average for CI_2 with the ODE results for CI_2. Repeat again using a run count of 100. How do the simulation times compare as run counts increase? Do your ODE and SSA simulations agree as you increase the number of runs? Does ODE simulation appear to be reasonably accurate for this example?

4.9   In this problem, use a tool such as `iBioSim` that supports both ODE and SSA simulation. Simulate the reaction-based model from Problem 1.9 using ODE simulation with a time limit of 2100 and print interval of 50, and note the simulation time. Repeat using SSA simulation and a run count of 1. Again, note the simulation time. Select an important species, and plot the results for both the ODE and SSA simulations. Repeat the SSA simulation using a run count of 10, and plot the average for this important species with its ODE results. Repeat again using a run count of 100. How do the simulation times compare as run counts increase? Do your ODE and SSA simulations agree as you increase the number of runs? Does ODE simulation appear to be reasonably accurate for this example?

4.10   Implement Gillespie's SSA using your favorite programming language.

4.11   Implement Gibson and Bruck's next reaction method using your favorite programming language.

4.12   Create an SPN model for the reaction-based model from Problem 1.3.

4.13   Create an SPN model for the reaction-based model from Problem 1.4.

4.14   Create an SPN model for the reaction-based model from Problem 1.5.

4.15   Create an SPN model for the reaction-based model from Problem 1.6.

4.16   Create an SPN model for the reaction-based model from Problem 1.7.

4.17   Create an SPN model for a portion of the reaction-based model from Problem 1.8.

4.18   Create an SPN model for a portion of the reaction-based model from Problem 1.9.

Chapter 5

Reaction-Based Abstraction

The grand aim of all science is to cover the greatest number of empirical facts by logical deduction from the smallest number of hypotheses or axioms.

—Albert Einstein

There is no abstract art. You must always start with something. Afterward you can remove all traces of reality.

—Pablo Picasso

As described in Chapter 4, several techniques have been proposed to reduce the computational costs of stochastic simulations. Ultimately, however, these methods are still limited in the size of the models that they can analyze. Given the substantial computational requirements for numerical simulation of even modest size genetic circuits, model abstraction is absolutely essential. To reduce the cost of simulation, this chapter describes methods to simplify the original reaction-based model by applying several *reaction-based abstractions*. These abstractions reduce the model's size by removing irrelevant or rapid reactions. Each abstraction examines the structure of the reaction-based model and, whenever possible, it applies transformations to simplify the model. The result is a new abstracted reaction-based model with less reactions and species which often substantially lowers the cost of simulation by not only reducing the model size but also eliminating many fast reactions that slow down simulation. The reduced model is also easier to intuitively visualize crucial components and interactions.

The remainder of this chapter describes each of the reaction-based abstractions. Section 5.1 presents *irrelevant node elimination*. Section 5.2 describes various *enzymatic approximations*. Section 5.3 presents *operator site reduction*. An alternative reduction for operator sites that uses *statistical thermodynamics* is presented in Section 5.4. Section 5.5 describes *dimerization reduction*. Section 5.6 shows how the reaction-based abstractions reduce the complexity of the analysis for the phage λ model. Finally, Section 5.7 describes *stoichiometry amplification*.

5.1   Irrelevant Node Elimination

In a large system, there may be some species that do not have significant influence on the species of interest, S_i. In such cases, simulation time can be reduced by removing such irrelevant species and reactions. Irrelevant node elimination uses reachability analysis on the model to detect nodes that do not influence the species in S_i. As an example, consider the portion of the model for phage λ shown in Figure 5.1(a). Let us assume that the only species in S_i is CI. Therefore, the production and degradation reactions for CI, r_5 and r_1, are relevant which makes S_1 relevant as well as r_3, the reaction that creates it. Making r_3 relevant makes RNAP, P_{RE}, and CII relevant. Since CII is relevant, its production and degradation reactions, r_6 and r_2, are now relevant which makes S_2 relevant as well as r_4, the reaction that creates it. Making r_4 relevant makes P_R relevant. At this point, species O and its degradation reaction, r_{11}, are the only remaining elements that have not been marked as relevant. Therefore, they are irrelevant to the amount of CI and can be removed as shown in Figure 5.1(b).

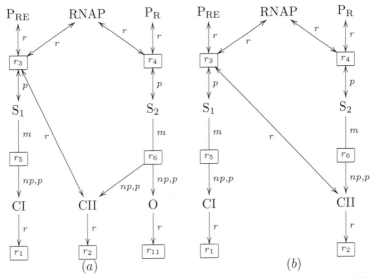

FIGURE 5.1:   Irrelevant node elimination: (a) original model, and (b) after reduction.

5.2 Enzymatic Approximations

A common motif found in biochemical systems are enzymatic reactions. Enzymatic reactions transform a substrate into a product catalyzed by an enzyme. Consider the following elementary enzymatic reactions where E is an enzyme, S is a substrate, C is a complex form of E and S, and P is a product.

$$E + S \underset{k_{-1}}{\overset{k_1}{\rightleftharpoons}} C \overset{k_2}{\longrightarrow} E + P$$

This model includes four species and three reactions, and it is depicted graphically in Figure 5.2(a). Using the law of mass action, the following reaction rate equations can be derived:

$$\frac{d[C]}{dt} = k_1[E][S] - k_{-1}[C] - k_2[C] \qquad (5.1)$$

$$\frac{d[P]}{dt} = k_2[C] \qquad (5.2)$$

When the complex C dissociates into E and S much faster than it is converted into the product P (i.e., $k_{-1} \gg k_2$), simulation time is dominated by this unproductive reaction. The *Production-Passage-Time Approximation* (PPTA) removes this unproductive reaction by approximating the passage time of C leading to P production. This reduced model, now with only two reactions, is shown below and in Figure 5.2(b).

$$E + S \overset{k_1'}{\longrightarrow} C \overset{k_2}{\longrightarrow} E + P$$

where $k_1' = \frac{k_1 k_2}{k_{-1} + k_2}$. In other words, the rate of C production is reduced to only produce those molecules of C that are destined to be converted into P.

When the total enzyme concentration, $[E_t]$, is much less than the substrate concentration plus the *Michaelis-Menten constant* (i.e., $[E_t] \ll [S] + K_M$ where $K_M = (k_{-1} + k_2)/k_1$), the enzymatic reaction network can be reduced even further by making a *steady-state assumption*. In particular, the steady-state concentration of the enzyme-substrate complex, $[C]$, is assumed to arrive very quickly. It is then assumed that in the steady-state, $[C]$ does not change appreciably (i.e., $d[C]/dt \approx 0$). Using this assumption, Equation 5.1 becomes:

$$[C] = \frac{k_1[E][S]}{k_{-1} + k_2} \qquad (5.3)$$

The concentration of free enzyme, $[E]$, can be rewritten in terms of the total concentration of enzyme, $[E_t]$, and the concentration of enzyme-substrate complex, $[C]$, as follows:

$$[E] = [E_t] - [C] \qquad (5.4)$$

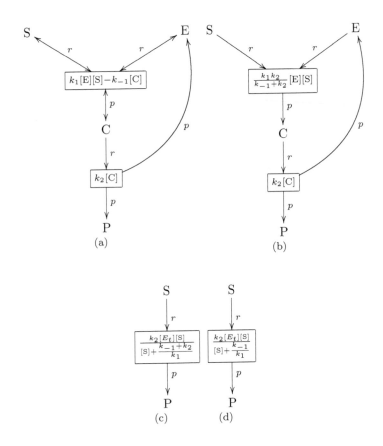

FIGURE 5.2:   Enzymatic approximations: (a) original enzymatic model,
(b) model after production-passage-time approximation,
(c) model after quasi-steady-state approximation, and
(d) model after rapid equilibrium approximation.

Substituting Equation 5.4 into Equation 5.3, after rearrangement, results in:

$$[C] = \frac{[E_t][S]}{[S] + \frac{k_{-1}+k_2}{k_1}} \tag{5.5}$$

Substituting Equation 5.5 into Equation 5.2 results in:

$$\frac{d[P]}{dt} = \frac{k_2[E_t][S]}{[S] + \frac{k_{-1}+k_2}{k_1}} = \frac{V_{max}[S]}{[S] + K_M} \tag{5.6}$$

where $V_{max} = k_2[E_t]$ and $K_M = (k_{-1} + k_2)/k_1$. This gives us the *Michealis-Menten* equation form shown below:

$$S \xrightarrow{\frac{V_{max}[S]}{[S]+K_M}} P \tag{5.7}$$

This equation is also known as the *quasi-steady-state approximation*, and it reduces the model to only two species and one reaction as shown in Figure 5.2(c).

In the case where k_2 is the rate limiting step (i.e., $k_{-1} \gg k_2$), this equation can be simplified even further as shown below and in Figure 5.2(d):

$$\frac{d[P]}{dt} = \frac{V_{max}[S]}{[S] + \frac{k_{-1}}{k_1}} \tag{5.8}$$

In this case, it is known as the *rapid equilibrium approximation*.

Using these enzymatic approximations, a reaction-based model can be reduced by searching for patterns matching the enzymatic reaction template shown in Figure 5.2(a). This procedure considers each species E as a potential enzyme, and each of the following must be true:

1. Species E must not be in S_i, and it must be a reactant in at least one reaction r_1.

2. Reaction r_1 must be reversible, have two reactants, a kinetic law of the form: $k_f[E][S] - k_r[C]$, and species C must not be in S_i.

3. The initial concentration of the complex species C must be zero.

4. Species C must be a product of only one reaction, r_1, reactant in only one reaction, r_2, and modifier in no reactions.

5. Reaction r_2 must not be reversible and have only one reactant and no modifiers.

6. Reaction r_2 must have only one or two products including species E, and it must have a kinetic law of the form: $k_2[C]$.

When such a pattern is found, the model is updated using one of the enzymatic approximations just described. Which approximation is applied depends on which assumptions on the ratios on rates and initial conditions are met. Note that enzyme E may be a reactant in multiple reactions known as *competitive enzymatic reactions*. In this case, for each reaction, a configuration is formed that includes the substrate S, complex C, equilibrium constant $K_1 = k_f/k_r$, production rate k_2, complex forming reaction r_1, and product forming reaction r_2. After finding all configurations for a given enzyme E, the abstraction procedure loops through the set of configurations to form an expression that is used in the denominator in each new rate law as well as forming a list of all the substrates that bind to the enzyme E. Next, for each configuration $(S, C, K_1, k_2, r_1, r_2)$, it makes the substrate S a reactant for r_2, makes all other substrates modifiers for r_2, creates a new rate law for r_2, and removes species C and reaction r_1. Finally, the enzyme E is removed. The application of enzymatic approximations not only reduces the size of the model, but also improves simulation time by removing fast reactions.

Figure 5.3(a) shows a graphical representation of a competitive enzymatic reaction from the phage λ model in which the proteins CII and CIII compete to bind to the protease P1 to produce the complexes P1·CII and P1·CIII, respectively. In the complex form, the protease degrades CII and CIII. The result of the rapid equilibrium approximation is shown in Figure 5.3(b) in which the protease, its complex forms, and the reactions that form these complexes are removed.

5.3   Operator Site Reduction

Genetic circuits often include many operator sites to which transcription factors bind. The operator sites are similar to enzymes in that their numbers are typically small relative to the numbers of RNAP and transcription factor molecules, and the binding and unbinding of these transcription factors to the operator sites is typically rapid as compared with the rate of open complex formation. Using these observations, a similar approximation method called *operator site reduction* can be derived. This method merges reactions while removing operator sites and their complexes from a reaction-based model.

For example, the rate of production of a protein may be inhibited by a repressor molecule. In other words, as the amount of a repressor increases, the rate of protein production decreases. A repressor typically binds to an operator site to prevent RNAP from binding to a promoter to start transcription. However, other mechanisms exist with similar dynamical behavior. In many situations, it may take multiple repressor molecules to inhibit production. For example, in phage λ, it takes four molecules of CI (i.e., two dimers) to repress

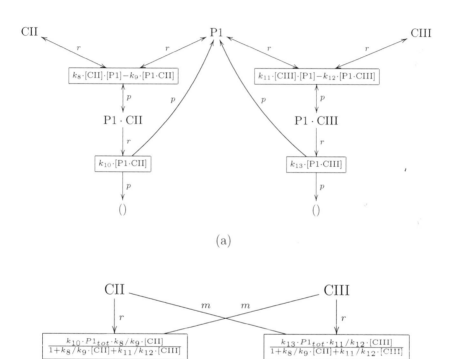

(a)

(b)

FIGURE 5.3: Competitive enzymatic reaction example: (a) original model, and (b) abstracted model (courtesy of Kuwahara *et al.* (2006)).

Cro production. Therefore, repression of this form can be modeled by the following reactions:

$$nR \underset{k_{-1}}{\overset{k_1}{\rightleftharpoons}} R_n$$

$$R_n + O \underset{k_{-2}}{\overset{k_2}{\rightleftharpoons}} R_nO$$

where R is the repressor, and O is the operator site to which the repressor complex, R_n, binds. Using the law of mass action, the following rate equations can be derived:

$$\frac{d[R]}{dt} = n(k_{-1}[R_n] - k_1[R]^n) \tag{5.9}$$

$$\frac{d[O]}{dt} = k_{-2}[R_nO] - k_2[R_n][O] \tag{5.10}$$

Assuming that the reactions are rapid and therefore always in equilibrium (i.e., $\frac{d[R]}{dt} = \frac{d[O]}{dt} \approx 0$), the following equations can be derived from Equations 5.9 and 5.10:

$$[R_n] = K_1[R]^n \tag{5.11}$$
$$[R_nO] = K_2[R_n][O] \tag{5.12}$$

where $K_1 = k_1/k_{-1}$ and $K_2 = k_2/k_{-2}$. Let us also assume that the concentrations of the intermediate complexes $R_1, \ldots, R_{n-1}$ are negligible. In other words, the repressor is complete or not formed at all. Last, let us assume that the concentration of operator sites, $[O]$, is small when compared with the total concentration of repressor molecules, $[R_t]$. Using these assumptions and Equations 5.11 and 5.12, the total concentration of operator sites is:

$$[O_t] = [O] + [R_nO] = [O](1 + K_1K_2[R]^n) \tag{5.13}$$

Rearranging Equation 5.13 results in the proportion of operator sites that are free of repressor as a function of repressor concentration:

$$f([R]) = \frac{[O]}{[O_t]} = \frac{1}{1 + K_1K_2[R]^n} \tag{5.14}$$

Note that $f([R])$ decreases monotonically as $[R]$ increases as shown in Figure 5.4. This function is known as a *Hill function* where $1/\sqrt[n]{K_1K_2}$ is the value of $[R]$ where $f([R])$ is $1/2$. The Hill function is a type of *sigmoid function* in which the steepness of the transition region around this point increases as the value of n increases. Indeed, in the limit as n approaches infinity, the Hill function becomes a step function.

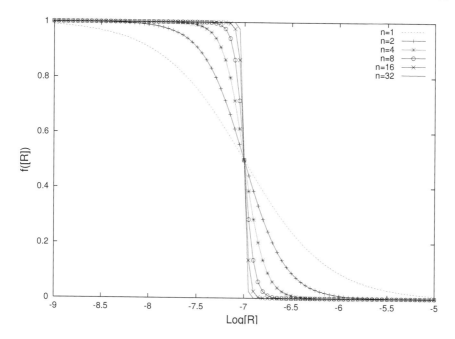

FIGURE 5.4:   The fraction of operator sites free of repressor as a function of repressor concentration.

As another example, let us consider n molecules of activator A binding to an operator O to enhance production of a protein. This is represented using the following reactions:

$$nA \xrightleftharpoons[k_{-1}]{k_1} A_n$$

$$A_n + O \xrightleftharpoons[k_{-2}]{k_2} A_nO$$

Using the law of mass action, the rates of change for A and O are represented using the following differential equations:

$$\frac{d[A]}{dt} = n(k_{-1}[A_n] - k_1[A]^n) \tag{5.15}$$

$$\frac{d[O]}{dt} = k_{-2}[A_nO] - k_2[A_n][O] \tag{5.16}$$

Again, assuming that the reactions are rapid (i.e., $\frac{d[A]}{dt} = \frac{d[O]}{dt} \approx 0$) results in the following equations:

$$[A_n] = K_1[A]^n \tag{5.17}$$

$$[A_nO] = K_2[A_n][O] \tag{5.18}$$

where $K_1 = k_1/k_{-1}$ and $K_2 = k_2/k_{-2}$. Again, let us assume that the concentrations of $A_1, \ldots, A_{n-1}$ are negligible and that $[O] << [A_t]$ (total concentration of activator). Using these assumptions and Equations 5.17 and 5.18 results in the following relationships:

$$[O] = [O_t] - [A_n O] \tag{5.19}$$

$$[A_n O] = K_1 K_2 [A]^n ([O_t] - [A_n O]) \tag{5.20}$$

$$f([A]) = \frac{[A_n O]}{[O_t]} = \frac{K_1 K_2 [A]^n}{1 + K_1 K_2 [A]^n} \tag{5.21}$$

Again, this equation takes the form of a Hill function. In this case, the proportion of activated operator sites increases monotonically as the concentration of A increases as shown in Figure 5.5, and it again behaves more like a step function as the value of n increases.

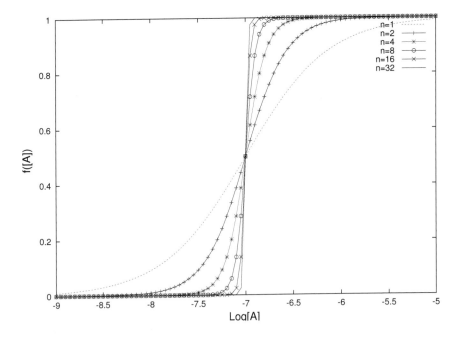

FIGURE 5.5: The fraction of operator sites bound to an activator as a function of activator concentration.

Putting it all together, let us consider an operator site, O, which can be repressed by a repressor, R, preventing the production of a protein, P. Let us also assume that the production rate is at a low basal rate, k_b, until enhanced by an activator, A, to a higher activated rate, k_a. The reactions for this operator site are shown below:

$$O + R \underset{k_{-1}}{\overset{k_1}{\rightleftharpoons}} O \cdot R$$

$$O + \text{RNAP} \underset{k_{-2}}{\overset{k_2}{\rightleftharpoons}} O \cdot \text{RNAP} \xrightarrow{k_b} O \cdot \text{RNAP} + np \; P$$

$$O + \text{RNAP} + A \underset{k_{-3}}{\overset{k_3}{\rightleftharpoons}} O \cdot \text{RNAP} \cdot A \xrightarrow{k_a} O \cdot \text{RNAP} \cdot A + np \; P$$

Using the law of mass action, some of the differential equations describing the behavior of this system are shown below:

$$\frac{d[O \cdot R]}{dt} = k_1[O][R] - k_{-1}[O \cdot R] \tag{5.22}$$

$$\frac{d[O \cdot \text{RNAP}]}{dt} = k_2[O][\text{RNAP}] - k_{-2}[O \cdot \text{RNAP}] \tag{5.23}$$

$$\frac{d[O \cdot \text{RNAP} \cdot A]}{dt} = k_3[O][\text{RNAP}][A] - k_{-3}[O \cdot \text{RNAP} \cdot A] \tag{5.24}$$

$$\frac{d[P]}{dt} = np \; k_b[O \cdot \text{RNAP}] + np \; k_a[O \cdot \text{RNAP} \cdot A] \tag{5.25}$$

Assuming that the binding to operator sites is rapid and therefore always in equilibrium (i.e., assuming $\frac{d[O \cdot R]}{dt} = \frac{d[O \cdot \text{RNAP}]}{dt} = \frac{d[O \cdot \text{RNAP} \cdot A]}{dt} \approx 0$), Equations 5.22, 5.23, and 5.24 reduce to:

$$[O \cdot R] = K_1[O][R] \tag{5.26}$$

$$[O \cdot \text{RNAP}] = K_2[O][\text{RNAP}] \tag{5.27}$$

$$[O \cdot \text{RNAP} \cdot A] = K_3[O][\text{RNAP}][A] \tag{5.28}$$

where $K_1 = k_1/k_{-1}$, $K_2 = k_2/k_{-2}$, and $K_3 = k_3/k_{-3}$. Using Equations 5.26, 5.27, and 5.28, the total concentration of operator sites, $[O_t]$, can be expressed as follows:

$$[O_t] = [O] + [O \cdot R] + [O \cdot \text{RNAP}] + [O \cdot \text{RNAP} \cdot A] \tag{5.29}$$

$$= [O] \left(1 + K_1[R] + K_2[\text{RNAP}] + K_3[\text{RNAP}][A] \right) \tag{5.30}$$

Rearranging Equation 5.30 results in the free operator site concentration as a function of the total operator concentration:

$$[O] = \frac{[O_t]}{1 + K_1[R] + K_2[\text{RNAP}] + K_3[\text{RNAP}][A]} \tag{5.31}$$

Substituting Equation 5.31 into Equation 5.27 and Equation 5.28 followed by substituting the results into Equation 5.25 produces the model after operator site reduction:

$$\frac{d[P]}{dt} = \frac{np \; (k_b K_2[\text{RNAP}] + k_a K_3[\text{RNAP}][A]) \, [O_t]}{1 + K_1[R] + K_2[\text{RNAP}] + K_3[\text{RNAP}][A]} \tag{5.32}$$

The first step of operator site reduction is to identify operators within the reaction-based model. This identification is accomplished by assuming that an operator is a species small in number that is neither produced nor degraded. Each species O is considered in turn as a potential operator site, and the following conditions are checked:

1. O's initial concentration must not be greater than a threshold.

2. Species O must not be in $\mathbf{S_i}$, and it must be a reactant in at least one reaction r_1.

3. Reaction r_1 must be reversible, have two or more reactants, exactly one product, and a kinetic law of the form: $k_f \cdot f([s_1], \ldots, [s_n]) - k_r[C]$.

4. The operator complex C must not be in $\mathbf{S_i}$, and it must be the product of one reaction and reactant of no reactions.

5. In each r_2 that C appears as a modifier, there must be no reactants, no other modifiers, one product, and a kinetic law of the form: $k_o[C]$.

For each reaction r_1, the algorithm creates a configuration $K_i X_i$ where $K_i = k_{fi}/k_{ri}$, and X_i is the product of reactant concentrations for r_1 excluding O.

Suppose our algorithm has identified an operator O, and there are N configurations in which transcription factors and RNAP can bind to it. For each species O identified as an operator, the reaction-based model is transformed using the following steps:

1. Create a sum: $Z = 1 + \sum_{j=1}^{N} K_j X_j$.

2. For each configuration i, create a new reaction which has as modifiers the reactants of the corresponding complex formation reaction r_1.

3. The kinetic law for this reaction is:

$$\frac{k_o O_0 K_i X_i}{Z}$$

4. Assuming O_0 is the total amount of operator, then $K_i X_i / Z$ is the proportion of O_0 in the i-th configuration.

The reaction-based model for the P_{RE} promoter from the phage λ decision circuit is shown graphically in Figure 5.6(a). This model includes two reversible reactions that bind $RNAP$ and CII to P_{RE}, and two reactions that each can produce np molecules of the protein CI. The operator sites associated with P_{RE} can be in one of three possible configurations. In particular, they can be empty, bound to RNAP alone (i.e., S_1), or bound to both RNAP and CII (i.e., S_2). Assuming that rates of the operator site binding and unbinding reactions are much faster than the rates of open complex formation, our method can apply operator site reduction. The result is the reduced model shown in Figure 5.6(b) which has only three species and two reactions. The abstracted model essentially determines the probability that operator sites associated with P_{RE} are in a configuration that results in the production of CI. This abstraction eliminates the rapid binding and unbinding reactions of transcription factors and RNAP to the operator sites that slow down simulation.

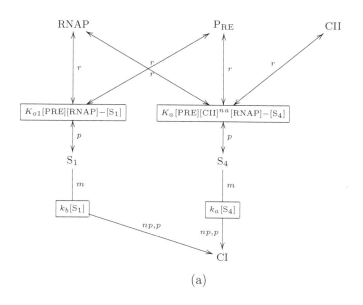

(a)

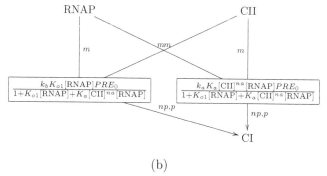

(b)

FIGURE 5.6: Operator site reduction: (a) original model, and (b) abstracted model.

The two reactions in the model shown in Figure 5.6(b) have the same reactants, modifiers, and products, so they can be combined using another reduction known as *similar reaction combination*. The resulting reduced model with only one reaction is shown in Figure 5.7(a). After performing operator site reduction to all the operators, RNAP only appears as a modifier in every reaction. In this case, another abstraction method known as *modifier constant propagation* can be applied. The method replaces each instance of [RNAP] with the constant RNAP_0 (i.e., the initial concentration of RNAP) as shown in Figure 5.7(b).

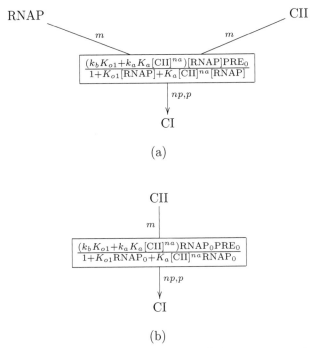

FIGURE 5.7: Similar reaction and modifier constant reductions:
(a) model after similar reaction combination, and
(b) model after modifier constant propagation.

5.4 Statistical Thermodynamical Model

Similar to operator site reduction, the *statistical thermodynamical model* proposed by Shea and Ackers also removes the rapid reactions involved in binding and unbinding of transcription factors to operator sites. This model incorporates rules and assumptions from previous genetic, biochemical, and structural studies and combines this with statistical thermodynamic assumptions. In particular, this model assumes that occupancy of operator sites can be determined by equilibrium statistical thermodynamic probabilities. Therefore, from this model, the probability of each potential configuration of transcription factors and RNAP bound to the operator sites can be determined. In other words, the model would not include specific reactions for the binding of molecules to operator sites, but instead during each simulation cycle, it determines the configuration of the operator sites using a probability distribution given by this thermodynamic model.

As an example, consider the potential configurations of the CI_2 molecule bound to the O_R operator. Some of the potential configurations are shown in Figure 5.8. Namely, after CI dimerizes, CI_2 can potentially bind to any one of the three O_R operator sites that is empty as shown in Figure 5.8(a). Recall that CI_2 bound to adjacent operator sites interact. Therefore, when CI_2 is bound to all three operator sites, the CI_2 molecules on O_R1 and O_R2 bind cooperatively as shown in Figure 5.8(b). It is assumed that cooperative interaction between O_R2 and O_R3 only happens in the rare case when O_R1 is vacant as shown in Figure 5.8(c). Recall that when O_R1 or O_R2 are occupied, the P_R promoter is blocked meaning that the *cro* gene is off. When O_R3 is occupied, the P_{RM} promoter is blocked meaning that the *cI* gene is off. Another assumption for this model is that in mutants in which one operator is damaged that the other operators work as before.

Using the assumptions just described, there are eight potential configurations of the O_R operator which are shown in Table 5.1. A free energy contribution, ΔG_s, relative to the reference (s=0) is assigned to each configuration. The values ΔG_1, ΔG_2, and ΔG_3 are the intrinsic free energies for the binding of CI_2 to the operator sites, and they are related to the equilibrium constants presented earlier by the equation:

$$\Delta G_i = -RT \ln K_i \qquad (5.33)$$

The values ΔG_{12} and ΔG_{23} are the free energy contributions due to cooperativity. The values for these free energy parameters can be determined from experimental results as described below.

Using the principles of statistical thermodynamics, the probability of the O_R operator being in each configuration, s, as a function of CI concentration

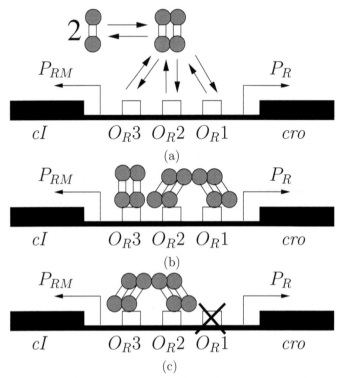

FIGURE 5.8: Configurations of CI_2 molecules on the O_R operator. (a) After dimerizing, CI can bind to any O_R operator site. (b) When CI_2 is bound to all three operator sites, both P_R and P_{RM} are repressed. Note that CI_2's bound to O_R2 and O_R1 bind cooperatively. (c) CI_2's can also bind cooperatively when bound to O_R3 and O_R2 when O_R1 is empty or mutated.

TABLE 5.1: Free energies for O_R configurations with CI_2 (Ackers *et al.*, 1982)

s	O_R3	O_R2	O_R1	Free energy contributions	ΔG_s
0	—	—	—	Reference	0
1	—	—	CI_2	ΔG_1	-11.7
2	—	CI_2	—	ΔG_2	-10.1
3	CI_2	—	—	ΔG_3	-10.1
4	—	CI_2^*	CI_2	$\Delta G_1 + \Delta G_2 + \Delta G_{12}$	-23.8
5	CI_2	—	CI_2	$\Delta G_1 + \Delta G_3$	-21.8
6	CI_2^*	CI_2	—	$\Delta G_2 + \Delta G_3 + \Delta G_{23}$	-22.2
7	CI_2	CI_2^*	CI_2	$\Delta G_1 + \Delta G_2 + \Delta G_3 + \Delta G_{12}$	-33.9

*Indicates that adjacent CI_2 molecules bind cooperatively.

can be expressed by:

$$f_s = \frac{exp(-\Delta G_s/RT)[\text{CI}_2]^{i(s)}}{\sum_s exp(-\Delta G_s/RT)[\text{CI}_2]^{i(s)}} \tag{5.34}$$

where ΔG_s is the free energy associated with configuration s from Table 5.1, $[\text{CI}_2]$ is the concentration of unbound CI_2 molecules, and $i(s)$ is the number of CI_2 molecules bound to O_R in configuration s. These probabilities can be combined to determine the probability of each operator site containing a CI_2 molecule using the equations below:

$$f_{OR1} = f_1 + f_4 + f_5 + f_7 \tag{5.35}$$
$$f_{OR2} = f_2 + f_4 + f_6 + f_7 \tag{5.36}$$
$$f_{OR3} = f_3 + f_5 + f_6 + f_7 \tag{5.37}$$

The last equation is also the probability that the promoter P_{RM} is repressed by the presence of a CI_2 molecule in O_R3 and blocking RNAP from binding. The promoter P_R is repressed by the presence of a CI_2 molecule at either O_R1 or O_R2, so the probability of it being repressed is given by this equation:

$$f_{PR} = f_1 + f_2 + f_4 + f_5 + f_6 + f_7 \tag{5.38}$$

Using the equations just presented, it is possible to derive the free energy values used in Table 5.1 (i.e., ΔG_1, ΔG_2, ΔG_3, ΔG_{12}, and ΔG_{23}). The data shown in Table 5.2 are taken from DNase protection experiments (Johnson *et al.*, 1979). This data shows the concentration of CI_2 necessary in order to find each operator site occupied half of the time. The experiments were performed on the wild-type and five mutants in which various operator sites are removed via mutation. Using Equation 5.35, 5.36, and 5.37 and the 11 data points to provide values for $[\text{CI}_2]$ in Table 5.2, 11 simultaneous equations with five unknowns (the free energies) can be created of the form $f_{site} = 0.5$. Using a nonlinear least squares procedure, the five free energies found are shown in Table 5.3.

TABLE 5.2: Values for $[\text{CI}_2]$ for half occupation (units of 3nM) (Johnson *et al.*, 1979)

DNA Template	O_R3	O_R2	O_R1
O_R^+ (wild type)	25	2	1
O_R1^-	5	5	—
O_R2^-	25	—	2
O_R1^-, O_R2^-	25	—	—
O_R1^-, O_R3^-	—	25	—
O_R3^-	—	2	1

TABLE 5.3: Resolved interaction free energies for O_R (Ackers *et al.*, 1982)

		Energy, kcal
Individual site binding	ΔG_1	-11.69 ± 0.03
	ΔG_2	-10.10 ± 0.05
	ΔG_3	-10.09 ± 0.02
Cooperative interaction	ΔG_{12}	-1.99 ± 0.06
	ΔG_{23}	-1.94 ± 0.06

Figure 5.9 shows the probability of repression versus the concentration of CI_2. The first curve is the probability of repression of P_R as determined using Equation 5.38. The second curve is the probability of repression of P_R neglecting cooperativity. The final curve is the probability of repression of P_{RM} as determined by Equation 5.37. These results indicate that it takes about 25 times more CI to half repress P_{RM} than P_R. Also, cooperativity makes the repression of P_R more switch-like which helps maintain a stable lysogen while it allows for induction.

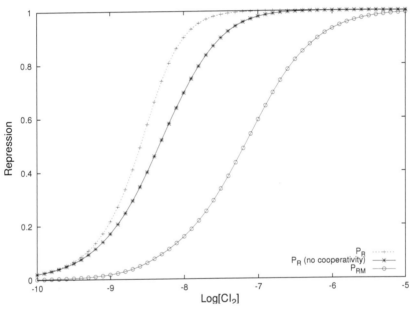

FIGURE 5.9: Probability of repression of the P_R and P_{RM} promoters versus $[CI_2]$.

Cro$_2$ and RNAP can also bind to the O_R operator sites. While Cro$_2$ can bind to individual operator sites, RNAP either binds to O_R3 or to both O_R2 and O_R1. There are a total of 40 potential configurations of the O_R operator sites with the free energies shown in Table 5.4. This table also shows the open complex rate of each configuration for P_R and P_{RM} (i.e., $k_{P_R}(s)$ and $k_{P_{RM}}(s)$). Using the following equations, the average open complex rate for P_R and P_{RM} can be calculated for various concentrations of CI$_2$ and Cro$_2$.

$$f_s = \frac{exp(-\Delta G_s/RT)[\text{CI}_2]^{i(s)}[\text{Cro}_2]^{j(s)}[\text{RNAP}]^{l(s)}}{\sum_s exp(-\Delta G_s/RT)[\text{CI}_2]^{i(s)}[\text{Cro}_2]^{j(s)}[\text{RNAP}]^{l(s)}} \quad (5.39)$$

$$k_{P_R} = \sum_s k_{P_R}(s)f_s \quad (5.40)$$

$$k_{P_{RM}} = \sum_s k_{P_{RM}}(s)f_s \quad (5.41)$$

The result is shown in Figure 5.10. These plots show that P_R is active initially when there is no CI$_2$ or Cro$_2$ while P_{RM} only becomes highly active after some CI$_2$ has been produced. In both cases, as the concentrations of CI$_2$ and Cro$_2$ increase, these promoters are strongly repressed.

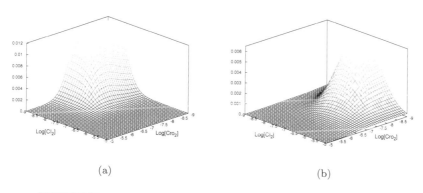

(a)   (b)

FIGURE 5.10:   Open complex rate for (a) P_R and (b) P_{RM}
versus [CI$_2$] and [Cro$_2$].

TABLE 5.4:   Free energies for O_R configurations (Shea and Ackers, 1985)

	State			ΔG_s	$k_{P_R}(s)$	$k_{P_{RM}}(s)$
s	O_R3	O_R2	O_R1	(kcal mol^{-1})	(sec^{-1})	(sec^{-1})
Non-liganded species						
1	—	—	—	0	0.0	0.0
Singly liganded species						
2	—	—	CI_2	-11.7	0.0	0.0
3	—	CI_2	—	-10.1	0.0	0.0
4	CI_2	—	—	-10.1	0.0	0.0
5	—	—	Cro_2	-10.8	0.0	0.0
6	—	Cro_2	—	-10.8	0.0	0.0
7	Cro_2	—	—	-12.1	0.0	0.0
8	RNAP	—	—	-11.5	0.0	0.001
9	—	RNAP		-12.5	0.014	0.0
Doubly liganded species						
10	—	CI_2^*	CI_2	-23.7	0.0	0.0
11	CI_2	—	CI_2	-21.8	0.0	0.0
12	CI_2^*	CI_2	—	-22.2	0.0	0.0
13	—	Cro_2	Cro_2	-21.6	0.0	0.0
14	Cro_2	—	Cro_2	-22.9	0.0	0.0
15	Cro_2	Cro_2	—	-22.9	0.0	0.0
16	RNAP	RNAP		-24.0	0.014	0.001
17	—	Cro_2	CI_2	-22.5	0.0	0.0
18	—	CI_2	Cro_2	-20.9	0.0	0.0
19	CI_2	—	Cro_2	-20.9	0.0	0.0
20	Cro_2	—	CI_2	-23.8	0.0	0.0
21	CI_2	Cro_2	—	-20.9	0.0	0.0
22	Cro_2	CI_2	—	-22.2	0.0	0.0
23	CI_2	RNAP		-22.6	0.014	0.0
24	RNAP	CI_2	—	-21.6	0.0	0.011
25	RNAP	—	CI_2	-23.2	0.0	0.001
26	Cro_2	RNAP		-24.6	0.014	0.0
27	RNAP	Cro_2	—	-22.3	0.0	0.001
28	RNAP	—	Cro_2	-22.3	0.0	0.001
Triply liganded species						
29	CI_2	CI_2^*	CI_2	-33.8	0.0	0.0
30	Cro_2	Cro_2	Cro_2	-33.7	0.0	0.0
31	Cro_2	CI_2^*	CI_2	-35.8	0.0	0.0
32	CI_2	Cro_2	CI_2	-32.6	0.0	0.0
33	CI_2^*	CI_2	Cro_2	-33.0	0.0	0.0
34	CI_2	Cro_2	Cro_2	-33.7	0.0	0.0
35	Cro_2	CI_2	Cro_2	-33.0	0.0	0.0
36	Cro_2	Cro_2	CI_2	-34.6	0.0	0.0
37	RNAP	CI_2^*	CI_2	-35.2	0.0	0.011
38	RNAP	Cro_2	Cro_2	-33.1	0.0	0.001
39	RNAP	Cro_2	CI_2	-34.0	0.0	0.001
40	RNAP	CI_2	Cro_2	-32.4	0.0	0.011

*Indicates that adjacent CI_2 molecules bind cooperatively.

5.5 Dimerization Reduction

Dimerization is another very rapid reaction which would be useful to remove whenever possible. A dimerization reaction has the following form:

$$2S_m \overset{k_+}{\underset{k_-}{\rightleftharpoons}} S_d$$

If S_m and S_d are in equilibrium, then the following is true:

$$[S_d] = K_d[S_m]^2 \tag{5.42}$$

where K_d is the equilibrium constant for dimerization (i.e., $K_d = k_+/k_-$). A simple dimerization reduction is to use this equation and replace all instances of $[S_d]$ with $K_d[S_m]^2$. A more accurate dimerization reduction method modifies the model to use total concentration, $[S_t]$, in place of the dimer concentration, $[S_d]$, and monomer concentration, $[S_m]$. The total concentration, $[S_t]$, can be expressed as follows:

$$[S_t] = [S_m] + 2[S_d] \tag{5.43}$$

Combining Equations 5.42 and 5.43 results in the following equation:

$$K_d[S_t]^2 - (4K_d[S_t] + 1)[S_d] + 4K_d[S_d]^2 = 0 \tag{5.44}$$

Solving Equation 5.44 allows us to express $[S_m]$ and $[S_d]$ in terms of $[S_t]$ as follows:

$$[S_d] = \frac{[S_t]}{2} - \frac{1}{8K_d}\left(\sqrt{8K_d[S_t] + 1} - 1\right) \tag{5.45}$$

$$[S_m] = \frac{1}{4K_d}\left(\sqrt{8K_d[S_t] + 1} - 1\right) \tag{5.46}$$

To perform the dimerization reduction, each reaction, r, is considered as a potential dimerization reaction. Reaction r is a dimerization reaction that can be reduced if the following conditions are true:

1. Reaction r must have one reactant, one product, no modifiers, and a kinetic law of the form: $k_+[S_m]^2 - k_-[S_d]$.

2. The monomer form, S_m, must not appear as a modifier in any reaction, and the dimer form, S_d, must not appear as a product in any reaction other than the dimerization reaction, r.

When a dimerization reaction is found, the reaction-based model is transformed as follows:

1. Create a species S_t with $[S_t]_0 = [S_m]_0 + 2[S_d]_0$.

2. In all reactions with S_m as a reactant, replace $[S_m]$ in the kinetic law with the following:

$$\frac{1}{4K_d}\left(\sqrt{8K_d[S_t]+1}-1\right)$$

3. In all reactions with S_m as a product, replace with S_t.

4. In all reactions with S_d as a reactant or modifier, replace $[S_d]$ in the kinetic law with the following:

$$\frac{[S_t]}{2} - \frac{1}{8K_d}\left(\sqrt{8K_d[S_t]+1}-1\right)$$

As an example, consider the reactions for CI from part of the phage λ decision circuit model shown in Figure 5.11(a). This species can only degrade in monomer form, but it must first dimerize before it can act as a transcription factor. The reduced model using the simple dimerization reduction is shown in Figure 5.11(b). Using the complete dimerization reduction, CI and CI$_2$ are replaced with CI$_t$, and the dimerization reaction is removed as shown in Figure 5.11(c).

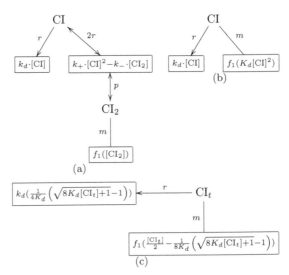

FIGURE 5.11: Dimerization reduction: (a) original model, (b) abstracted model using the simple dimerization reduction, and (c) abstracted model using the complete dimerization reduction.

5.6   Phage λ Decision Circuit Example

After performing all the abstractions described to this point, the abstracted reaction-based model for the CI/CII portion of the phage λ decision circuit is reduced to just two species and four reactions as shown in Figure 5.12(a). Figure 5.12(b) shows the average result of 10,000 SSA runs for CI_total for this abstracted model as compared with the original model. The results are nearly the same, but the abstracted results are generated in less than 40 seconds while the original results required more than 20 minutes.

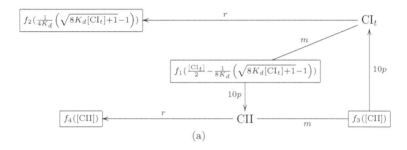

(a)

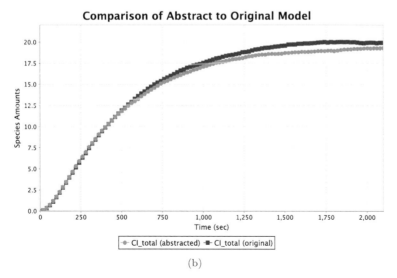

(b)

FIGURE 5.12:   (a) Abstracted reaction-based model for the CI/CII portion of the phage λ decision circuit. (b) Comparison of the average of CI_total for 10,000 SSA runs before and after reaction-based abstraction.

The complete reaction-based model for the phage λ decision circuit has 61 species and 75 reactions. After reaction-based abstraction, the size of the model is reduced to only 5 species and 11 reactions as shown in Figure 5.13(a). Both models are simulated for 10,000 SSA runs for one cell cycle (i.e., 2100 seconds). If the total number of CI molecules exceeds 328 before the number of Cro molecules exceeds 133, then the run is said to result in lysogeny. The simulations are run for MOIs ranging from 1 to 50. Figure 5.13(b) shows the probability of lysogeny for MOIs from 0 to 10 for both the original and abstracted model which shows that the results are nearly the same. The simulation of the original model takes 56.5 hours while the abstracted model is reduced to 9.8 hours.

5.7 Stoichiometry Amplification

After performing reaction-based abstractions, analysis can be greatly accelerated due to the reduction in model size, elimination of rapid reactions, and simplification of the rate laws. Stochastic simulation can be further accelerated using *stoichiometry amplification* which modifies the stoichiometry of all or part of the reactions by multiplying them by an amplification factor n and dividing the reaction rates by the same factor. This abstraction method is illustrated in Figure 5.14(a) and 5.14(b). While this modification has no impact on differential equation analysis, it can substantially reduce the cost of stochastic simulation. In particular, it reduces the propensities allowing for the system to advance faster by taking larger time steps. As another example, Figure 5.14(a) shows the result of performing stoichiometry amplification using a factor of 10 on the degradation reactions in the CI/CII portion of the phage λ decision circuit model. In this example, there is an added benefit of a reduction in the state space size. While in general every species count can take on any arbitrary value, in this reduced model, the species counts can only be multiples of ten.

5.8 Sources

The enzymatic reaction formulation was initially proposed in Henri (1903) while the Michaelis-Menten approximation was first presented in Michaelis and Menten (1913). The theoretical basis to the Michaelis-Menten approximation is given in Briggs and Haldane (1925) which is further justified using perturbation theory in Segel and Slemrod (1989). In Rao and Arkin (2003),

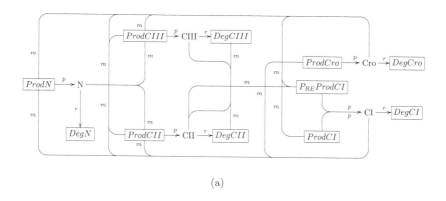

(a)

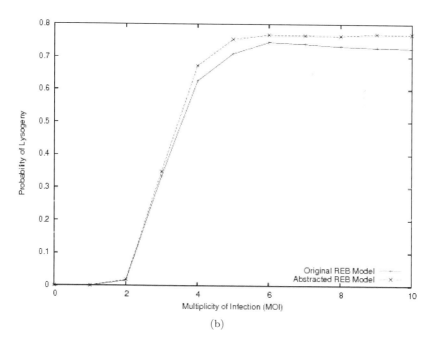

(b)

FIGURE 5.13: (a) Abstracted reaction-based model for the complete phage λ decision circuit. (b) Comparison of results of the probability of lysogeny versus multiplicity of infection using 10,000 SSA runs before and after abstraction. Results before abstraction require 56.5 hours of simulation time while those after abstraction require only 9.8 hours (courtesy of Kuwahara *et al.* (2006)).

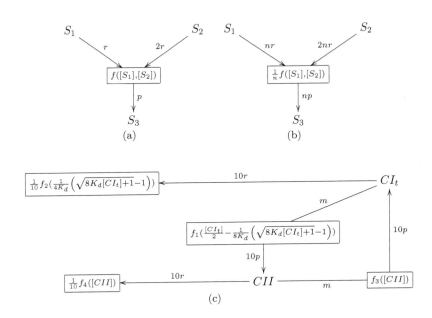

FIGURE 5.14: Stoichiometry amplification: (a) original model, and (b) model after stoichiometry amplification. (c) CI/CII portion of the phage λ decision circuit model after stoichiometry amplification with a factor of 10 on degradation rates.

the quasi-steady-state assumption is utilized to remove fast reactions to simplify analysis using a modified version of the SSA. The production-passage-time approximation is developed in Kuwahara and Myers (2008). Reducing operator site models using the fraction of operator sites in various configurations is proposed in Tyson and Othmer (1978). The statistical thermodynamical model presented in Section 5.4 was originally presented in Ackers *et al.* (1982) and later extended in Shea and Ackers (1985). It has been utilized in various stochastic analyses such as those in Arkin *et al.* (1998); Wolf and Arkin (2002). The theoretical basis to the dimerization reduction first appeared in Santillán and Mackey (2004). The reaction-based abstraction methods described in this chapter as well as the phage λ analysis are from Kuwahara *et al.* (2006); Kuwahara (2007).

Problems

5.1 Below are the chemical reactions involved in a competitive enzymatic reaction in which two substrates compete for a single enzyme.

$$E + S_1 \overset{k_1}{\underset{k_{-1}}{\rightleftharpoons}} ES_1 \overset{k_2}{\rightarrow} E + P_1$$

$$E + S_2 \overset{k_3}{\underset{k_{-3}}{\rightleftharpoons}} ES_2 \overset{k_4}{\rightarrow} E + P_2$$

5.1.1. Using the law of mass action, write down the equations for the rates of change of $[S_1]$, $[S_2]$, $[ES_1]$, $[ES_2]$, $[P_1]$, and $[P_2]$.

5.1.2. If the amount of enzyme, E, is much less than either of the substrates, S_1 and S_2, then a steady-state approximation can be used to simplify the equations for the rates of P_1 and P_2 formation. Using this approximation, derive an equation for $\frac{d[P_1]}{dt}$ in terms of the concentrations of the substrates, $[S_1]$ and $[S_2]$, the total enzyme concentration $[E_t]$, and the rate constants. You may also assume that $k_2 << k_{-1}$ and that $k_4 << k_{-3}$ to further simplify the derivations (hint: use this assumption at the beginning to simplify the rate equations for $[ES_1]$ and $[ES_2]$).

5.2 The first chemical reaction below is a representation for transcription and translation of a protein P. This process begins when RNAP binds to an operator/promoter site O and forms a complex C_1. At this point, either the RNAP can fall off O or transcription and translation can be initiated resulting in the protein P. The second chemical reaction represents that the protein R can bind to O forming the complex C_2 which represses transcription blocking the production of P.

$$O + \text{RNAP} \overset{k_1}{\underset{k_{-1}}{\rightleftharpoons}} C_1 \overset{k_2}{\rightarrow} O + \text{RNAP} + P$$

$$O + R \overset{k_3}{\underset{k_{-3}}{\rightleftharpoons}} C_2$$

5.2.1. Using the law of mass action, write down the equations for the rates of change of $[P]$, $[C_1]$, and $[C_2]$.

5.2.2. Write an equation for the total operator concentration, $[O_t]$.

5.2.3. The reactions above can be thought of being like an enzymatic reaction where O acts like the enzyme. If the amount of O is much less than that of RNAP or R, then a steady-state approximation can be used to simplify the equations for the rate of P production. Using this approximation, derive an equation for $\frac{d[P]}{dt}$ in terms of the concentrations of $[\text{RNAP}]$, $[R]$, and $[O_t]$, and the rate constants. You may also assume that $k_2 << k_{-1}$ to further simplify the derivations (hint: solve for $[C_2]$ in terms of $[O]$ and $[R]$ first followed by using your $[O_t]$ equation to solve for $[O]$ in terms of $[O_t]$, $[C_1]$, and $[R]$).

5.2.4. Assuming that $[O_t] = 1$ nM and $[RNAP] = 30$ nM, at what concentration of $[R]$ is the rate of production of $[P]$ reduced by one half.

5.3 The P_{RE} promoter has the following four configurations:

s	O_1	O_2	Free energy
1	—	—	Reference
2	CII	—	ΔG_1
3	—	RNAP	ΔG_2
4	CII	RNAP	$\Delta G_1 + \Delta G_2 + \Delta G_{12}$

where ΔG_{12} represents the cooperative effect of CII bound to O_1 drawing RNAP to O_2. Assume the following experimental data:

DNA Template	O_1	O_2
wild type	12	30
O_1-	—	60
O_2-	85	—

where each entry is the number of nM before there is half occupation of the corresponding operator sites. For wild type, the data is the number of nM until half of the P_{RE} promoters have configuration 4. Using this experimental data, determine values for ΔG_1, ΔG_2, and ΔG_{12}. For RT, assume a value of 0.5961. (hint: Use O_1- and O_2- data first to calculate ΔG_2 and ΔG_1, respectively. Note that for these mutants not all configurations are possible. Also, note that the concentrations above are in nanomoles. Next, use the wild type data to find ΔG_{12}.)

5.4 The P_L promoter has the following ten configurations:

s	O_1	O_2	Free energy
1	—	—	Reference
2	Cro_2	—	ΔG_1
3	—	Cro_2	ΔG_2
4	CI_2	—	ΔG_3
5	—	CI_2	ΔG_4
6	—	RNAP	ΔG_5
7	Cro_2	Cro_2	$\Delta G_1 + \Delta G_2$
8	Cro_2	CI_2	$\Delta G_1 + \Delta G_4$
9	CI_2	Cro_2	$\Delta G_2 + \Delta G_3$
10	CI_2	CI_2	$\Delta G_3 + \Delta G_4 + \Delta G_{34}$

where ΔG_{34} represents the cooperative effect of CI_2 bound to O_1 drawing another CI_2 to O_2. Assume the following experimental data:

DNA Template	Protein	O_1	O_2
wild type	CI_2	1.6	2.3
O_1-	CI_2	—	44
O_1-	Cro_2	—	1.5
O_2-	CI_2	3	—
O_2-	Cro_2	11.4	—
O_2-	$RNAP_2$	0.8	—

where each entry is the number of nM before there is half occupancy of the corresponding operator sites by the protein that is provided. Using this experimental data, determine values for ΔG_1, ΔG_2, ΔG_3, ΔG_4, ΔG_5, and ΔG_{34}. For RT, assume a value of 0.5961. (hint: Use O_1- and O_2- data first to calculate ΔG_1, ΔG_2, ΔG_3, ΔG_4, and ΔG_5. Note that for these mutants not all configurations are possible. Also, note that the concentrations above are in nanomoles. Next, use the wild type data to find ΔG_{34}.)

5.5 The chemical reactions below represent transcription and translation to produce a protein P. Transcription is initiated by the activator protein A which must first dimerize before acting as a transcription factor.

$$2A \underset{k_2}{\overset{k_1}{\rightleftharpoons}} A_2$$

$$O + A_2 + RNAP \underset{k_4}{\overset{k_3}{\rightleftharpoons}} C \overset{k_5}{\rightarrow} O + A_2 + RNAP + P$$

5.5.1. Using the law of mass action, write down the differential equation for $[A_2]$ considering only the first reaction.

5.5.2. Make a steady state assumption (i.e., $d[A_2]/dt \approx 0$), and solve for $[A_2]$ in terms of $[A]$.

5.5.3. Using the law of mass action, write down the differential equations for $[P]$ and $[C]$ considering only the second reaction.

5.5.4. Write an equation for the total operator concentration, $[O_t]$.

5.5.5. Make a steady state assumption (i.e., $d[C]/dt \approx 0$), and solve for $[C]$ in terms of $[O_t]$, $[A_2]$, $[RNAP]$, and the rate constants (hint: use your equation for $[O_t]$ to eliminate $[O]$).

5.5.6. Using your results, write an equation for $d[P]/dt$ in terms of $[O_t]$, $[A]$, $[RNAP]$, and the rate constants (hint: be sure to use your equation for $[A_2]$).

5.5.7. Create an SBML model for both the original and abstracted models in iBioSim or your favorite SBML editor. Simulate both models using an ODE simulator for 1.0 seconds with a time step of 0.01 seconds starting with initial concentrations of (O, RNAP, A, A_2, C, P) $= (1, 30, 10, 0, 0, 0)$ and assuming $k_1 = 1$, $k_2 = 0.1$, $k_3 = 1$, $k_4 = 1$, and $k_5 = 0.1$. How does the results from the abstracted model compare with the full model.

5.6   For the reactions below, find a reduced model using abstraction. To simplify the result, use the simple dimerization abstraction in which the dimer species is rewritten in terms of the monomer species using Equation 5.42.

$$2A \underset{k_2}{\overset{k_1}{\rightleftarrows}} A_2$$

$$2R \underset{k_3}{\overset{k_2}{\rightleftarrows}} R_2$$

$$O + R_2 \underset{k_5}{\overset{k_4}{\rightleftarrows}} C1$$

$$O + \text{RNAP} \underset{k_7}{\overset{k_6}{\rightleftarrows}} C2 \overset{k_8}{\rightarrow} O + \text{RNAP} + P$$

$$O + A_2 + \text{RNAP} \underset{k_{10}}{\overset{k_9}{\rightleftarrows}} C3 \overset{k_{11}}{\rightarrow} O + A_2 + \text{RNAP} + P$$

5.7   Use abstraction to reduce the reaction-based model from Problem 1.3 and determine the simplified ODE model. To simplify the result, use the simple dimerization abstraction in which the dimer species is rewritten in terms of the monomer species using Equation 5.42.

5.8   Use abstraction to reduce the reaction-based model from Problem 1.4 and determine the simplified ODE model. To simplify the result, use the simpler dimerization abstraction in which the dimer species is rewritten in terms of the monomer species using Equation 5.42.

5.9   Use abstraction to reduce the reaction-based model from Problem 1.5 and determine the simplified ODE model. To simplify the result, use the simpler dimerization abstraction in which the dimer species is rewritten in terms of the monomer species using Equation 5.42.

5.10   Use abstraction to reduce the reaction-based model from Problem 1.6 and determine the simplified ODE model. To simplify the result, use the simpler dimerization abstraction in which the dimer species is rewritten in terms of the monomer species using Equation 5.42.

5.11   Use abstraction to reduce the reaction-based model from Problem 1.7 and determine the simplified ODE model. To simplify the result, use the simpler dimerization abstraction in which the dimer species is rewritten in terms of the monomer species using Equation 5.42.

5.12   Use iBioSim to abstract the reaction-based model from Problem 1.8.

5.13   Use iBioSim to abstract the reaction-based model from Problem 1.9.

Chapter 6

Logical Abstraction

> Somebody who thinks logically is a nice contrast to the real world.
>
> —The Law of Thumb

Electrical engineers have vast experience in modeling and analyzing circuits and systems composed of thousands or even millions of interconnected complex components. One of the most important electrical circuits is the microprocessor. Figure 6.1(a) shows the first microprocessor, Intel's 4004, which was introduced in 1971. This processor is composed of 2,300 transistors, the basic element of an integrated electrical circuit, and it operates at 108 kilohertz. After 30 years of technology improvements, the complexity of electrical circuits that can be designed has grown at a incredible rate. The circuit shown in Figure 6.1(b) is the Intel®Pentium®4 Processor which was introduced in 2000. This processor is composed of 42 million transistors, and it operates at 1.5 gigahertz. The increase in performance over the 4004 is simply amazing. If the speed of automobiles had improved at a similar rate during that 30 year period, one could now drive from San Francisco to New York in only 13 seconds! One reason that we can design circuits such as these is that methods of logical abstraction and analysis help us to deal with this complexity.

We are fortunate that the complexity of the biochemical circuits that we wish to analyze are not increasing, but their initial complexity is staggering. An example biochemical circuit, the reactions involved in *E. coli* metabolism, is shown in Figure 6.2. While this level of complexity may be manageable, if the regulatory reactions are added then the figure would appear black. In order to analyze circuits such as these, model abstraction is going to be absolutely essential. The question that this chapter addresses is whether the logical abstraction applied to electrical circuits can be applied successfully to biochemical circuits.

As mentioned earlier, the regulation of genetic circuits is often controlled by Hill functions. In the limit, these Hill functions become step functions which can be encoded logically as described in Section 6.1. Using this *logical encoding*, one can construct *piecewise models* which are presented in Section 6.2. Section 6.3 presents *stochastic finite-state machines* that can be analyzed using *Markov chain analysis* as described in Section 6.4. Finally, Section 6.5 presents *qualitative logical models*.

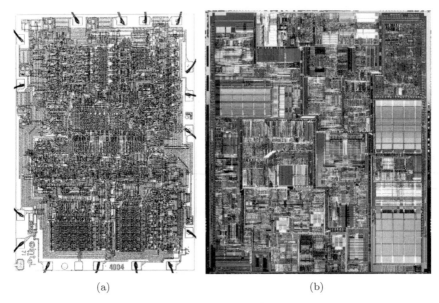

(a)   (b)

FIGURE 6.1: Example electrical circuits (courtesy of the Intel Museum). (a) Intel 4004 microprocessor. (b) Intel®Pentium®4 microprocessor.

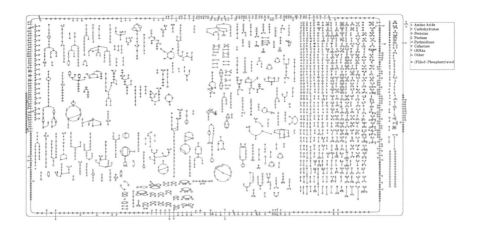

FIGURE 6.2: All reactions involved in *E. coli* metabolism (courtesy of http://www.ecocyc.org).

6.1 Logical Encoding

Electrical circuits are often classified as being either *analog* (i.e., having continuous valued states) or *digital* (i.e., having discrete valued states). While analog circuits must be analyzed using expensive differential equation simulation, digital circuits are analyzed using substantially more efficient *switch-level simulation* in which each wire is assumed to take only one of two binary values. It should be noted though that digital circuits are actually also analog circuits, but a logical abstraction has been employed to reduce the complexity of analysis. This logical abstraction is essential as complex integrated circuits, such as microprocessors, cannot be efficiently analyzed using differential equation analysis. Can the efficiency of genetic circuit analysis also be improved using such a logical abstraction?

As described in the last chapter, the effects of inhibition or activation in genetic circuits can often be modeled with Hill functions such as in the form shown below:

$$\frac{1}{(1 + K_j x_j^n)} \quad \text{OR} \quad \frac{K_j x_j^n}{(1 + K_j x_j^n)} \tag{6.1}$$

where $\theta_j = \sqrt[n]{a/(K_j - aK_j)}$ is the *critical threshold* where the change occurs, and a is an amplifier in the range of $[0.5, 1.0]$. As n increases, the time spent in the transition region decreases relative to the response time of the system. In other words, this function begins to behave like a step function with a step change at θ_j. In this case, x_j could be encoded using a binary variable which is false when $x_j < \theta_j$, and it is true when $x_j \geq \theta_j$.

When a species is used as a transcription factor at several locations, it is likely that its critical threshold is different at each location. For example, CI_2 represses both the P_R and P_L promoters while also activating its own production at P_{RM}. Each of these effects occur at different critical threshold levels. A single species may also affect the activity of a promoter differently at various concentration levels. For example, the activity level of the P_{RM} promoter changes twice as CI_2 concentration increases as shown in Figure 6.3. Namely, at low levels, there is minimal activity until CI_2 reaches a moderate level and activates its own production. Finally, at high levels, it represses its own production. Clearly, CI_2 cannot be represented accurately using a single binary variable. Therefore, it is often the case that species need to be encoded using *n-ary* rather than just binary levels.

Assume that species x_j has N_j critical thresholds $\theta_j^1, \ldots, \theta_j^i, \ldots, \theta_j^{N_j}$ that satisfy:

$$\theta_j^0 < \theta_j^1 < \ldots < \theta_j^i < \ldots < \theta_j^{N_j} < \theta_j^{N_j+1}$$

where $\theta_j^0 = 0$ and $\theta_j^{N_j+1} = \infty$. Using these critical thresholds, the states

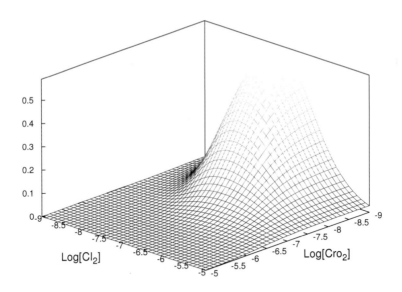

FIGURE 6.3: Activity of the P_{RM} promoter.

of x_j can be partitioned into *critical intervals* $(A_j^0, A_j^1, \ldots, A_j^{N_j})$ where $A_j^i = [\theta_j^i, \theta_j^{i+1})$. An n-ary variable b_j is created which can take any value in $\{0, 1, \ldots, N_j\}$. The initial value of b_j is the largest i such that $[x_j]_0 \geq \theta_j^i$ where $[x_j]_0$ is the initial concentration of x_j. The critical thresholds divide the space of potential species concentrations into n-dimensional *regulatory domains* that are separated by hyperplanes $x_j = \theta_j^i$. The total number of these n-dimensional domains is:

$$\prod_{j=1}^{n}(N_j + 1) \tag{6.2}$$

where n is the number of species. An example with three species with 18 3-dimensional domains is shown in Figure 6.4. An assignment to each of the n-ary variables, b_j, uniquely selects an individual domain.

As an example, let us consider an abstracted model for the CI/CII portion of the phage λ decision circuit shown below:

$$\frac{d[CI]}{dt} = \frac{np \; P_{RE}RNAP(k_bK_{o1} + k_aK_a[CII])}{1 + K_{o1}RNAP + K_aRNAP[CII]} - k_d[CI]$$

$$\frac{d[CII]}{dt} = \frac{np \; k_oP_RK_{o2}RNAP}{1 + K_{o2}RNAP + K_rK_d[CI]^2} - k_d[CII]$$

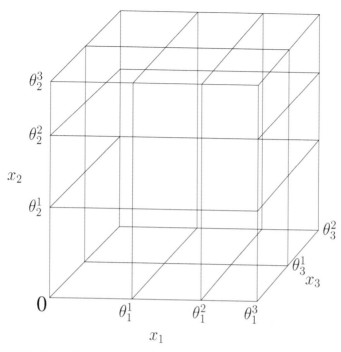

FIGURE 6.4: Example regulatory domains for an example with three species. Species x_1 and x_2 have three critical intervals while x_3 has two. Note that although $\theta_1^3 = \theta_2^3 = \theta_3^2 = \infty$, they are shown bounded in order to simplify the presentation.

These equations indicate the following critical thresholds and intervals assuming the amplifier, a, is 0.5:

$$\theta_{CI}^1 = \frac{1}{\sqrt{K_r K_d}} = 7, \ \theta_{CII}^1 = \frac{1}{K_a RNAP} = 21$$

$$A_{CI}^0 = [0,7), A_{CI}^1 = [7,\infty), \ A_{CII}^0 = [0,21), A_{CII}^1 = [21,\infty)$$

6.2 Piecewise Models

Within each regulatory domain, the rate of change of each species can be approximated by a *piecewise linear differential equation* (PLDE) of the form:

$$\frac{dx_j}{dt} = f_j(\mathbf{x}) - g_j(\mathbf{x})x_j, \quad j = 1,\dots,n \tag{6.3}$$

where f_j and g_j are piecewise constant functions and $\mathbf{x} = [x_1, \ldots, x_n]$ is a vector of species concentrations. Each function, f_j and g_j, changes value when a species concentration x_j crosses one of the θ_j^i thresholds. In other words, f_j and g_j can be rewritten into this form:

$$f_j(\mathbf{x}) = \sum_{l \in L} \alpha_{jl} B_{jl}(\mathbf{x}) \geq 0 \qquad (6.4)$$

where α_{jl} is a constant rate of change, and B_{jl} is an n-ary logical function that returns a binary value under the conditions that the rate of production of species x_j equals α_{jl}. The function B_{jl} is composed of a conjunction of terms of the form $(b_k = i)$. As an example, the PLDE model for the CI/CII portion of the phage λ decision circuit is given below:

$$\frac{d[CI]}{dt} = np \; P_{RE}(k_b + k_a(b_{CII} = 1)) - k_d[CI]$$

$$\frac{d[CII]}{dt} = np \; k_o P_R(b_{CI} = 0) - k_d[CII]$$

Inside each domain B, the behavior is linear and quite simple where B is defined by a Boolean formula of the form: $(b_1 = i_1) \wedge \ldots \wedge (b_n = i_n)$, and they are denoted by a state vector of the form $(i_1 \ldots i_n)$. For the example, the domains are (00), (01), (10), and (11). Let us denote the value of $\mathbf{f} = [f_1, \ldots, f_n]$ within one of the B domains, by the value $\mathbf{f}^B$, and $\mathbf{g} = [g_1, \ldots, g_n]$ is g^B. Therefore, within the B domain, the behavior of $\mathbf{x}$ reduces to the simple linear differential equation $\frac{d\mathbf{x}}{dt} = f^B - g^B \mathbf{x}$ which has the solution:

$$\mathbf{X}(t) = \Phi^B + (\mathbf{X}(t_0) - \Phi^B)e^{\gamma(t_0 - t)} \quad \text{where } \Phi^B = \mathbf{f}^B/\mathbf{g}^B \qquad (6.5)$$

For the example, the solutions are:

$$\Phi^{00} = \left(\frac{np \; k_o P_R}{k_d}, \frac{np \; k_b P_{RE}}{k_d} \right) = (19, 0.05)$$

$$\Phi^{01} = \left(0, \frac{np \; k_b P_{RE}}{k_d} \right) = (0, 0.05)$$

$$\Phi^{10} = \left(\frac{np \; k_o P_R}{k_d}, \frac{np \; P_{RE}(k_b + k_a)}{k_d} \right) = (19, 20)$$

$$\Phi^{11} = \left(0, \frac{np \; P_{RE}(k_b + k_a)}{k_d} \right) = (0, 20)$$

As t approaches $+\infty$, $\mathbf{X}(t)$ approaches Φ^B until $\mathbf{X}(t)$ reaches one of the boundaries of the B domain. If Φ^B is within the B domain, then $\mathbf{X}(t)$ reaches a stable stationary point at Φ^B. What happens to this function though when it encounters a domain boundary? Assuming that for species x_j that boundary B is bounded between θ_j^i and θ_j^{i+1}, then the behavior is defined as follows:

- If $\Phi^B < \theta_j^i$, then all trajectories in B that encounter the wall defined by $x_j = \theta_j^i$ are leaving the domain B.

- If $\Phi^B > \theta_j^{i+1}$, then all trajectories in B that encounter the wall defined by $x_j = \theta_j^{i+1}$ are leaving the domain B.

- If $\theta_j^i < \Phi^B < \theta_j^{i+1}$, then all trajectories that encounter the boundary defined by either $x_j = \theta_j^i$ or $x_j = \theta_j^{i+1}$ enter the domain B.

A boundary between two domains is *transparent* if trajectories enter one domain and leave the other domain through this boundary. The boundary is *black* if trajectories leave both domains from this boundary. Finally, a boundary is *white* if trajectories enter both domains from this boundary. If a boundary is black or white, the result is a *sliding motion*. If the boundary is black, then the solution proceeds along the boundary until it either reaches another boundary or a stable point on the boundary. If the boundary is white, the solution can either proceed sliding along the boundary or leave it at any point, since a white wall is unstable.

The behavior for the PLDE model of the CI/CII portion of the phage λ decision circuit can be depicted graphically using the flow graph shown in Figure 6.5. The solution, Φ^{00} resides within the (00) domain which means that it is a stable solution. This result corresponds with the differential equation simulation of a stable solution within this domain. The stochastic simulation result can also be seen here in that Φ^{00} is near the (10) boundary. If random production of CII causes the system to cross the boundary into (10) then CI production becomes enabled, and the system is driven towards Φ^{10}. This may either drive the system back into the (00) domain or the (11) domain. If the system enters the (11) domain, it is driven towards Φ^{11} which is in the (01) domain. If the system crosses into the (01) domain, then it is driven towards Φ^{01} which is in the (00) domain. In other words, stochastic behavior can cause the system to cycle through all the domains resulting in oscillatory behavior.

Chapter 4 described stochastic Petri nets as an alternative formalism for describing stochastic kinetic models. There is also a variant of Petri nets, *hybrid Petri nets* (HPNs), that can be used to represent piecewise differential equation models. HPNs include both a discrete part that can model discrete states such as the current regulatory domain that the system is in as well as a continuous part that can model continuous quantities like species concentrations. There are numerous ways that continuous quantities have been added to Petri nets. *Labeled hybrid Petri nets* (LHPNs) add the continuous values as auxiliary variables that evolve over time. These variables can be sampled in enabling conditions added to the transitions. Also, their rates of change can be modified by assignments on transitions. An example LHPN model for the CI/CII portion of the phage λ decision circuit is shown in Figure 6.6. Note that the rates are calculated based upon the assumption that $[CI]$ is 0 in places p_0 and p_1 while it is 7 in places p_2 and p_3. Similarly, $[CII]$ is

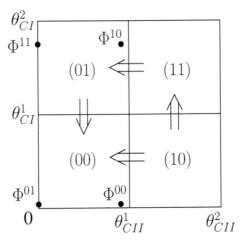

FIGURE 6.5: Flow graph for the PLDE model of the CI/CII portion of the phage λ decision circuit.

assumed to be 0 in places p_0 and p_3 while it is assumed to be 21 in places p_1 and p_2. In the initial state, P_0 is marked, CI and CII are both 0, and the rate of change of CI is initially 0.0004 while that for CII is 0.14. When CII reaches 23, transition t_0 becomes enabled. When t_0 fires, the rate of change of CI changes to 0.15 and CII changes to -0.02, and the marking moves to P_1. In this state, if CI reaches 7 first, then transition t_1 fires, but if CII decreases below 19 first, then transition t_4 fires. Note that the enabling conditions for t_0 and t_4 compare against different values to prevent t_0 and t_4 from firing continuously.

A piecewise model for the entire phage λ lysis-lysogeny decision circuit can also be constructed though it is more difficult to visualize since it requires five dimensions. Therefore, only a brief description of it is given here using the diagram shown in Figure 6.7. In this figure, Boolean logic gates have been added to represent the influence that each species has on each promoter. Note that these gates are *leaky* in that the logic only represents which states that the promoters are the most active. For example, the promoter P_L is the most active when both Cro and CI are at their lowest threshold level. In other words, the rate of production of N and CIII decreases as the values of b_{CI} and b_{Cro} increase. Similarly, the rate of production of CI from P_{RE} increases as the values of b_{CII} and b_{CIII} increase. The rates of production of CII and CIII increase as b_N increases as N closes the terminator switches. Finally, the rate of production of CI and Cro from P_{RM} and P_R is a bit more involved. Namely, the rate of CI production reduces as b_{Cro} increases while the rate of Cro production reduces as b_{CI} increases. However, while the rate of Cro production is high initially, the rate of CI production is low until b_{CI} reaches a moderate value.

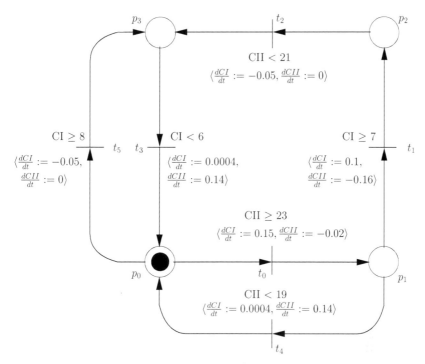

FIGURE 6.6: LHPN model for the CI/CII portion of the phage λ circuit. In the initial state, CI and CII are both 0, and the rate of change of CI is initially 0.0004 while that for CII is 0.14.

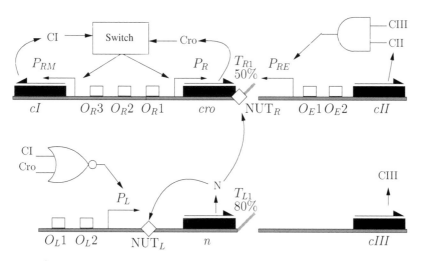

FIGURE 6.7: Piecewise model for the phage λ decision circuit.

6.3 Stochastic Finite-State Machines

The piecewise models just presented still track the exact concentrations of each species making their state space infinite. Alternatively, a stochastic finite-state machine (FSM) only tracks the n-ary encoding value for each species. This creates a purely logical representation of the genetic circuit. Analysis of a stochastic FSM can be accomplished using either stochastic simulation or Markov chain analysis which is described in the next section. A stochastic FSM can often be efficiently analyzed while maintaining the high-level quantitative behavior.

Before generating an FSM, the reaction-based model must be transformed such that all reactions have either one reactant *or* one product, but not both. This is often the case after applying the reaction-based abstractions. If this property does not hold, it can be made to hold using *reaction splitization*. One form of reaction splitization is called *multiple reactant/multiple product reaction splitization*, which splits an irreversible reaction with multiple reactants and multiple products into an irreversible reaction with multiple reactants and no products and an irreversible reaction with no reactants and multiple products. The second reaction has the original reactants changed to modifiers for the reaction. As an example, consider the reaction shown in Figure 6.8(a) that transforms species S_1 and S_2 into species S_3 and S_4 with a rate law $f([S_1], [S_2])$. After splitization, this reaction is transformed into the two reactions shown in Figure 6.8(b) with the same rate law. Next, *multiple reactant splitization* can be applied to the first reaction to produce two new reactions each with a single reactant with the other reactant as a modifier as shown in Figure 6.8(c). Finally, *multiple product splitization* can be applied to the second reaction to produce two new reactions each with a single product and the original reactants as modifiers as shown in Figure 6.8(d). The reaction rate equations derived from the original reaction and the split reactions are exactly the same, so there is no change in the differential equation analysis. For stochastic analysis, however, the split reaction model is an approximation.

A stochastic FSM is specified using a set of *guarded commands* that change the values of the n-ary variables. Each guarded command, $c_k \in C$, has a form:

$$G_k(\mathbf{b}) \xrightarrow{q_k} b_j := i$$

where the function $G_k(\mathbf{b})$ is the *guard*, q_k is the transition rate, and i is the n-ary value assigned to b_j as a result of c_k. A guard is a conjunction of literals of the form $(b_j = i)$. Each guarded command, c_k, is required to monotonically change the state of some variable in $\mathbf{b}$. Therefore, if b_j is assigned to i by c_k, then the guard must include a term of the form $(b_j = i - 1)$ or $(b_j = i + 1)$.

The guard for a reaction is derived from the n-ary variables for the species used in that reaction. As an example, let us consider the production of CI from the P_{RE} promoter and production of CII from the P_R promoter. This example

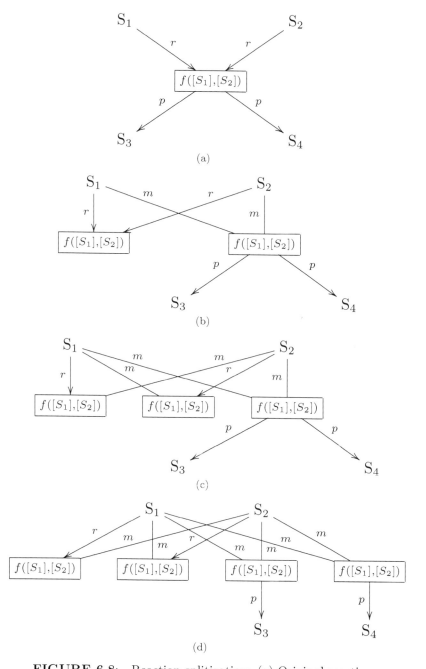

FIGURE 6.8: Reaction splitization: (a) Original reaction, (b) after multiple reactants/multiple products splitization, (c) after multiple reactants splitization, and (d) after multiple products splitization.

is derived from the complete phage λ model to make it more interesting. In particular, it is assumed that b_{CI} can take the values in $\{0, 1, 2\}$ and b_{CII} can take the values in $\{0, 1, 2, 3\}$. The resulting guarded commands for the production of CI are:

$$b_{CI} = 0 \wedge b_{CII} = 0 \xrightarrow{q_1} b_{CI} := 1$$
$$b_{CI} = 1 \wedge b_{CII} = 0 \xrightarrow{q_2} b_{CI} := 2$$
$$b_{CI} = 0 \wedge b_{CII} = 1 \xrightarrow{q_3} b_{CI} := 1$$
$$b_{CI} = 1 \wedge b_{CII} = 1 \xrightarrow{q_4} b_{CI} := 2$$
$$b_{CI} = 0 \wedge b_{CII} = 2 \xrightarrow{q_5} b_{CI} := 1$$
$$b_{CI} = 1 \wedge b_{CII} = 2 \xrightarrow{q_6} b_{CI} := 2$$
$$b_{CI} = 0 \wedge b_{CII} = 3 \xrightarrow{q_7} b_{CI} := 1$$
$$b_{CI} = 1 \wedge b_{CII} = 3 \xrightarrow{q_8} b_{CI} := 2$$

The guarded commands for the degradation of CI are:

$$b_{CI} = 2 \xrightarrow{q_9} b_{CI} := 1$$
$$b_{CI} = 1 \xrightarrow{q_{10}} b_{CI} := 0$$

The guarded commands for the production of CII are:

$$b_{CI} = 0 \wedge b_{CII} = 0 \xrightarrow{q_{11}} b_{CII} := 1$$
$$b_{CI} = 0 \wedge b_{CII} = 1 \xrightarrow{q_{12}} b_{CII} := 2$$
$$b_{CI} = 0 \wedge b_{CII} = 2 \xrightarrow{q_{13}} b_{CII} := 3$$
$$b_{CI} = 1 \wedge b_{CII} = 0 \xrightarrow{q_{14}} b_{CII} := 1$$
$$b_{CI} = 1 \wedge b_{CII} = 1 \xrightarrow{q_{15}} b_{CII} := 2$$
$$b_{CI} = 1 \wedge b_{CII} = 2 \xrightarrow{q_{16}} b_{CII} := 3$$
$$b_{CI} = 2 \wedge b_{CII} = 0 \xrightarrow{q_{17}} b_{CII} := 1$$
$$b_{CI} = 2 \wedge b_{CII} = 1 \xrightarrow{q_{18}} b_{CII} := 2$$
$$b_{CI} = 2 \wedge b_{CII} = 2 \xrightarrow{q_{19}} b_{CII} := 3$$

The guarded commands for the degradation of CII are:

$$b_{CII} = 3 \xrightarrow{q_{20}} b_{CII} := 2$$
$$b_{CII} = 2 \xrightarrow{q_{21}} b_{CII} := 1$$
$$b_{CII} = 1 \xrightarrow{q_{22}} b_{CII} := 0$$

The final step to generate an FSM is to assign a transition rate, q_k, to each guarded command. For each guarded command, c_k, that increases b_j to i:

$$q_k = \frac{m \cdot f_j(\Theta)}{\theta_j^i - \theta_j^{i-1}}$$

where m is the stoichiometry of s in the corresponding reaction, $f(\Theta)$ is the rate law for this reaction, and Θ is the critical levels that satisfy $G_k(\mathbf{b})$. For each guarded command, c_k, that decreases b_j to i:

$$q_k = \frac{m \cdot f_j(\Theta)}{\theta_j^{i+1} - \theta_j^i}$$

The transition rates for production of CI are:

$$q_1 = 10 \cdot f_3(0)/\theta_{CI}^1 = 0.000002$$
$$q_2 = 10 \cdot f_3(0)/(\theta_{CI}^2 - \theta_{CI}^1) = 0.000003$$
$$q_3 = 10 \cdot f_3(\theta_{CII}^1)/\theta_{CI}^1 = 0.00012$$
$$q_4 = 10 \cdot f_3(\theta_{CII}^1)/(\theta_{CI}^2 - \theta_{CI}^1) = 0.00016$$
$$q_5 = 10 \cdot f_3(\theta_{CII}^2)/\theta_{CI}^1 = 0.0014$$
$$q_6 = 10 \cdot f_3(\theta_{CII}^2)/(\theta_{CI}^2 - \theta_{CI}^1) = 0.0018$$
$$q_7 = 10 \cdot f_3(\theta_{CII}^3)/\theta_{CI}^1 = 0.0025$$
$$q_8 = 10 \cdot f_3(\theta_{CII}^3)/(\theta_{CI}^2 - \theta_{CI}^1) = 0.0032$$

The transition rates for degradation of CI are:

$$q_9 = f_2(\theta_{CI}^2)/(\theta_{CI}^2 - \theta_{CI}^1) = 0.0002$$
$$q_{10} = f_2(\theta_{CI}^1)/\theta_{CI}^1 = 0.00035$$

If the process being modeled is in the state $\mathbf{b}$, c_k can be executed if its guard is satisfied (i.e., $G_k(\mathbf{b})$ evaluates to true). The result of executing the guarded command is that a new state $\mathbf{b}'$ is reached in which $b_j' = i$ and $b_l' = b_l$ and for all $l \neq j$. The probability that c_k is executed is:

$$P(c_k) = G_k(\mathbf{b}) \cdot q_k \cdot \Delta t$$

where Δt is a time step which must be small enough such that the probability that two or more commands are executed in that time interval is negligible. The probability that no transition is taken in a Δt time step is:

$$(1 - (\textstyle\sum_{k=0}^{|C|} G_k(\mathbf{s}) \cdot q_k \cdot \Delta t))$$

A stochastic FSM can be analyzed using multiple stochastic simulation runs beginning in state $\mathbf{b_0}$. In each state, the simulation process determines whether or not to execute a guarded command in the next Δt time step using the probability functions defined above. If a guarded command is executed, the assignment associated with the command is performed resulting in a new state, and the transition probabilities are recalculated for this new state. This process continues until the desired simulation time has been reached. This simulation process though is inefficient, since as Δt is reduced to improve accuracy, the number of simulation steps that do not result in a state change increases significantly. Therefore, a more efficient approach is use a method such as the exact SSA which jumps to the time of the next state change.

6.4   Markov Chain Analysis

Stochastic FSMs can be efficiently analyzed using Markov chain analysis. Markov processes must satisfy the *Markov property*:

$$Pr[X(t+\tau) = y \mid X(s) = x(s), \forall s \leq t] =$$
$$Pr[X(t+\tau) = y \mid X(t) = x(t)], \quad \forall \tau > 0$$

In other words, the process is *memoryless* in that the probability of a future state depends only on the present state, and it does not depend on any past states or the amount of time in which it has already spent in the current state. A *homogeneous Markov process* also does not depend on the time t:

$$Pr[X(t+\tau) = y \mid X(t) = x] = Pr[X(\tau) = y \mid X(0) = x], \forall t, \tau > 0$$

Markov chains are Markov processes that have a discrete state space.

Discrete-time Markov chains (DTMC) have states that are observed only at discrete time points. Homogeneous DTMCs are specified by a *transition probability matrix*, P, composed of *single-step transition probabilities*:

$$p_{ij} = Pr[X_{n+1} = j \mid X_n = i], \text{ for all } n = 0, 1, \ldots$$

where $0 \leq p_{ij} \leq 1$ and $\sum_{\text{all } j} p_{ij} = 1$. As an example, let us consider the weather in Salt Lake City, Utah in January which can be snowy, overcast, or clear. The transition probability matrix for this example is:

$$P = \begin{pmatrix} 0.4 \ 0.4 \ 0.2 \\ 0.7 \ 0.3 \ 0.0 \\ 0.3 \ 0.2 \ 0.5 \end{pmatrix}$$

This matrix indicates, for example, that if today is snowy then tomorrow has a 40 percent chance of being snowy, 40 percent chance of being overcast, and a 20 percent chance of being clear. In other words, this matrix allows one to forecast tomorrow's weather based on today's weather.

In order to make longer range forecasts, one can construct n-step transition probabilities using the following recursive relationship:

$$P^n = PP^{(n-1)}$$

For example, to predict the weather in two days, one can simply square the matrix P to obtain:

$$P^2 = \begin{pmatrix} 0.5 \ \ 0.32 \ 0.18 \\ 0.49 \ 0.37 \ 0.14 \\ 0.41 \ 0.28 \ 0.31 \end{pmatrix}$$

This matrix indicates, for example, that if it is snowy today then there is a 50 percent chance of it being snowy in two days. If one multiplies the matrix P by itself many times (i.e., compute P^∞), eventually the current state has little impact on some state very distant in the future. In this case, all the rows of P^n become equal, and each row indicates the likelihood of a particular state in general. For our example, if only two significant digits are considered, this matrix is obtained after six steps:

$$P^6 = P^\infty = \begin{pmatrix} 0.48 \ 0.33 \ 0.19 \\ 0.48 \ 0.33 \ 0.19 \\ 0.48 \ 0.33 \ 0.19 \end{pmatrix}$$

This result means that knowing the current weather conditions gives you no additional information about the weather six days from now. This matrix also indicates that on an average day in January that there is a 48 percent chance of a snowy day, 33 percent chance of an overcast day, and a 19 percent chance of a clear day.

A state is *transient* when there is a non-zero probability that the DTMC will at some point never return to that state. A state is *recurrent* when the DTMC is guaranteed to return to this state at some point in the future. A state is *positive-recurrent* when its mean time to revisit is finite. A state is *null-recurrent* when its mean time to revisit is infinite. In a finite Markov chain, all states are transient or positive-recurrent. A state j is *periodic* with period p when upon leaving j it can only be returned to after a number of transitions that is a multiple of $p > 1$. A state with $p = 1$ is *aperiodic*. An *ergodic Markov chain* is positive-recurrent and aperiodic. A DTMC is *irreducible* if every state can be reached by every other state. A finite, aperiodic, irreducible Markov chain is ergodic.

One is often interested in determining the probability of being in a state:

$$\pi_i(n) = Pr[X_n = i]$$

The probability vector for all states is written as follows:

$$\pi(n) = [\pi_1(n), \pi_2(n), \ldots, \pi_i(n), \ldots]$$

The limit as n goes to ∞ is the *limiting distribution*:

$$\pi = \lim_{n \to \infty} \pi(n)$$

In an ergodic Markov chain, the limiting distribution is also known as a *steady-state distribution* and satisfies:

$$\pi = \pi P$$

To find the steady-state distribution, one can multiply P by itself repetitively as described earlier. This method can also be accelerated by squaring the matrix repetitively instead. This method, however, has several drawbacks. The main drawback is the time and memory requirements for matrix multiplication. In particular, P is typically a very large and sparse matrix (i.e., many entries are zero). However, after squaring the matrix, the zero entries tend to take non-zero values. If the matrix is kept sparse, then P can be represented with a sparse matrix data structure which is typically substantially smaller improving both runtime and memory usage. There have been numerous methods developed to find the steady-state distribution for Markov chains that address these issues. Two types of methods, *direct methods* and *iterative methods*, are briefly described below.

Direct methods solve the system of equations given by $\pi = \pi P$ using Gaussian elimination or other methods. Returning to our example, the following equation is setup in which s represents snowy, o represents overcast, and c represents clear:

$$[s \; o \; c] = [s \; o \; c] \begin{pmatrix} 0.4 \; 0.4 \; 0.2 \\ 0.7 \; 0.3 \; 0.0 \\ 0.3 \; 0.2 \; 0.5 \end{pmatrix}$$

Multiplying through results in the following:

$$[s \; o \; c] = [(0.4s + 0.7o + 0.3c)(0.4s + 0.3o + 0.2c)(0.2s + 0.5c)]$$

This represents three equations with three unknowns:

$$s = 0.4s + 0.7o + 0.3c$$
$$o = 0.4s + 0.3o + 0.2c$$
$$c = 0.2s + 0.5c$$

These equations can then be solved resulting in the following:

$$s = 0.48$$
$$o = 0.33$$
$$c = 0.19$$

These results match those obtained earlier.

Iterative methods apply the recursive formula $\pi(n) = \pi(n-1)P$ until *convergence* is achieved. Convergence can, at times, be slow and *periodicity* must be determined as described later. The convergence rate is also dependent on the initial state vector, $\pi(0)$. While the initial state vector is arbitrary, an initial guess that is closer to the actual solution converges faster. An application of the iterative method to our simple example is shown below:

$$\pi(1) = [1.0 \ \ 0.0 \ \ 0.0] \begin{pmatrix} 0.4 \ 0.4 \ 0.2 \\ 0.7 \ 0.3 \ 0.0 \\ 0.3 \ 0.2 \ 0.5 \end{pmatrix} = [0.4 \ \ 0.4 \ \ 0.2]$$

$$\pi(2) = [0.4 \ \ 0.4 \ \ 0.2] \begin{pmatrix} 0.4 \ 0.4 \ 0.2 \\ 0.7 \ 0.3 \ 0.0 \\ 0.3 \ 0.2 \ 0.5 \end{pmatrix} = [0.5 \ \ 0.32 \ \ 0.18]$$

$$\pi(3) = [0.5 \ \ 0.32 \ \ 0.18] \begin{pmatrix} 0.4 \ 0.4 \ 0.2 \\ 0.7 \ 0.3 \ 0.0 \\ 0.3 \ 0.2 \ 0.5 \end{pmatrix} = [0.48 \ \ 0.33 \ \ 0.19]$$

$$\pi(4) = [0.48 \ \ 0.33 \ \ 0.19] \begin{pmatrix} 0.4 \ 0.4 \ 0.2 \\ 0.7 \ 0.3 \ 0.0 \\ 0.3 \ 0.2 \ 0.5 \end{pmatrix} = [0.48 \ \ 0.33 \ \ 0.19]$$

When iterations are converging rapidly, one can check that the difference between two iterations is less than a desired accuracy, ϵ:

$$\|\pi(k) - \pi(k-1)\| < \epsilon$$

where $\|\mathbf{x}\| = \sqrt{x_1^2 + \ldots + x_n^2}$. When iterations are converging slowly, it is better to check the difference between two iterations separated by a number of steps, m, that is set based on the convergence rate:

$$\|\pi(k) - \pi(k-m)\| < \epsilon$$

Problems occur, however, when probabilities are small, so in this case, one should normalize before checking convergence as shown below:

$$\max_i \left(\frac{|\pi_i(k) - \pi_i(k-m)|}{|\pi_i(k)|} \right) < \epsilon$$

The period of an irreducible Markov chain is:

$$p = \gcd(l_1, \ldots, l_i, \ldots, l_c)$$

where gcd is the greatest common divisor, l_i is the length of the ith cycle in the Markov chain, and c is the total number of cycles. A *periodic Markov chain* (i.e., $p > 1$) does not converge for all m. As an example, let us assume that there are never two snowy days in a row and an overcast or clear day is always followed by a snowy day. The result of iterative analysis in this case is shown below:

$$\pi(1) = [1.0 \;\; 0.0 \;\; 0.0] \begin{pmatrix} 0.0 & 0.7 & 0.3 \\ 1.0 & 0.0 & 0.0 \\ 1.0 & 0.0 & 0.0 \end{pmatrix} = [0.0 \;\; 0.7 \;\; 0.3]$$

$$\pi(2) = [0.0 \;\; 0.7 \;\; 0.3] \begin{pmatrix} 0.0 & 0.7 & 0.3 \\ 1.0 & 0.0 & 0.0 \\ 1.0 & 0.0 & 0.0 \end{pmatrix} = [1.0 \;\; 0.0 \;\; 0.0]$$

$$\pi(3) = [1.0 \;\; 0.0 \;\; 0.0] \begin{pmatrix} 0.0 & 0.7 & 0.3 \\ 1.0 & 0.0 & 0.0 \\ 1.0 & 0.0 & 0.0 \end{pmatrix} = [0.0 \;\; 0.7 \;\; 0.3]$$

In this case, the iteration appears to never converge when two sequential steps are considered. A simple solution to deal with periodicity is to only check convergence after each p steps or some multiple of p steps. When convergence is detected, the steady state distributions from the last p steps are combined and then normalized by p. For our example, the result would be:

$$\pi = [1.0 \;\; 0.7 \;\; 0.3]/2 = [0.5 \;\; 0.35 \;\; 0.15]$$

Continuous-time Markov chains (CTMC), have states that can change at any arbitrary point in time. A stochastic FSM can be represented as a homogeneous CTMC. CTMCs are specified using a *transition rate matrix*, Q, rather than a transition probability matrix. Each entry in a transition rate matrix not only indicates how likely each next state is but also how soon that the system transitions to that state. The transition rate matrix for a homogeneous CTMC is defined as follows:

$$p_{ij}(\tau) = Pr[X(s + \tau) = j \mid X(s) = i]$$

$$q_{ij} = \lim_{\Delta t \to 0} \left\{ \frac{p_{ij}(\Delta t)}{\Delta t} \right\}, \;\; \text{for } i \neq j$$

$$q_{ii} = -\sum_{j \neq i} q_{ij}$$

Each non-diagonal q_{ij} entry represents a rate of change, so the larger the value, the faster the system moves to that state. The diagonal q_{ii} entries are negative since the likelihood of the system staying in the current state should diminish as time passes. For the weather example, a CTMC would describe the current weather at all points in time rather than for each discrete day. A transition rate matrix for our weather example is shown below:

$$Q = \begin{pmatrix} -6 & 4 & 2 \\ 4 & -4 & 0 \\ 4 & 4 & -8 \end{pmatrix}$$

One can find the steady-state distribution of a CTMC using its discrete-time *embedded Markov chain* (EMC) using the equation below:

$$\pi = \frac{-\phi D_Q^{-1}}{\|\phi D_Q^{-1}\|_1}$$

where ϕ is the steady-state distribution for the EMC, D_Q^{-1} is the inverse of the diagonal matrix of Q, and $\|\mathbf{x}\|_1 = \sum_{i=1}^{n} |x_i|$. The transition probability matrix for an EMC for Q is defined as follows:

$$s_{ij} = \begin{cases} \frac{q_{ij}}{\sum_{i \neq j} q_{ij}}, & i \neq j \\ 0 & i = j \end{cases}$$

Applying this method to our example results in the following EMC:

$$S = \begin{pmatrix} 0 & 0.67 & 0.33 \\ 1 & 0 & 0 \\ 0.5 & 0.5 & 0 \end{pmatrix}$$

The steady state distribution for S is:

$$\phi = (0.46, 0.39, 0.15)$$

The negative inverse of the diagonal matrix for Q is:

$$-D_Q^{-1} = \begin{pmatrix} 0.167 & 0 & 0 \\ 0 & 0.25 & 0 \\ 0 & 0 & 0.125 \end{pmatrix}$$

The result after multiplying this by ϕ is:

$$-\phi D_Q^{-1} = (0.08, 0.1, 0.02)$$

Summing all the entries of this vector results in:

$$\|\phi D_Q^{-1}\|_1 = 0.2$$

Putting these results together, the final steady state distribution is as follows:

$$\pi = \frac{-\phi D_Q^{-1}}{\|\phi D_Q^{-1}\|_1} = (0.4, 0.5, 0.1)$$

This result indicates that although a snowy day is the most likely state in the EMC that when the rates are considered, overcast becomes the most likely state. This is due to the fact that the overcast state is left slower (-4) than the snowy state (-6).

Figure 6.9(a) shows the CTMC for CI and CII from the complete phage λ decision circuit constructed from the guarded commands for the stochastic

FSM representation. Figure 6.9(b) is the EMC for this example annotated with the initial state probabilities. The transition probabilities for each state in the EMC are determined by dividing the transition rate by the sum of all transition rates exiting a state. Figure 6.9(c) is the result after 31 iterations of an iterative Markov chain analysis while Figure 6.9(d) is the result after 32 iterations. Notice that this example has a periodicity of two. Therefore, to compute the final steady state probability these two state probabilities must be summed together and divided by 2 to obtain the result shown in Figure 6.9(e). Next, each of these state probabilities is divided by the sum of all the transition rates exiting the state. Finally, these values must be normalized to obtain the steady state probability of the CTMC that is shown in Figure 6.9(f). This result indicates that the system is in the middle bottom two states 75 percent of the time (i.e., CI is high and CII is at a moderate concentration), and it is in one of the middle six states 98 percent of the time (i.e., CII stays at a moderate concentration while CI fluctuates).

Figure 6.10 compares both experimental and stochastic simulation data to results obtained using an iterative method on the CTMC model for the phage λ decision circuit. The results for the CTMC track the starved data points reasonably well. The CTMC results, however, are found in less than 7 minutes of computation time. Another important thing to note is that for the CTMC model, simulation results for the well-fed case can also be found in 7 minutes. These results could not be generated using the original model in reasonable time, since the number of simulation runs necessary is inversely proportional to the probability of lysogeny.

6.5   Qualitative Logical Models

All the models described so far require the estimation of binding affinities and kinetic parameters which are often difficult to obtain for genetic circuits. Studies have found, however, that these systems are often quite robust to parameter variation. From this observation, it may be possible to make reasonable behavioral predictions with only qualitative information. A *qualitative logical model* is similar to the stochastic FSM model except that no rate parameters are provided. While guarded commands can also be used for qualitative models, this section uses an n-ary logical equation representation which is a bit more compact. It is important to note that while a stochastic FSM may potentially enter any state, it would not be particularly informative if a qualitative logical model also could reach any state. Therefore, qualitative logical models typically only describe the most likely states and state transitions.

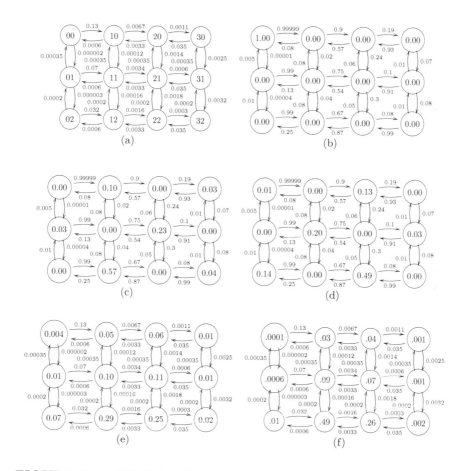

FIGURE 6.9: CTMC for CI and CII from the complete phage λ decision circuit model with CI increasing downwards and CII increasing to the right: (a) CTMC, (b) EMC with initial state probabilities, (c) state probabilities after 31 iterations, (d) state probabilities after 32 iterations, (e) steady-state distribution of the EMC, and (f) steady-state distribution of the CTMC.

Engineering Genetic Circuits

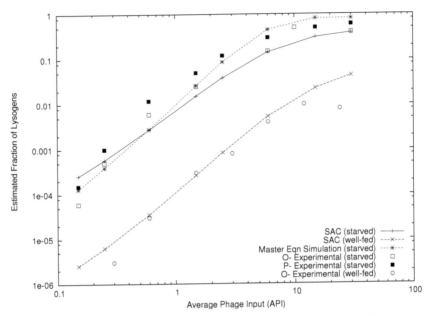

FIGURE 6.10: Comparison of CTMC results to experimental data (courtesy of Kuwahara *et al.* (2006)).

As an example, a qualitative logical model for the CI/CII portion of the phage λ decision circuit is shown below:

$$b_{CII} := (b_{CI} = 0)$$
$$b_{CI} := (b_{CII} = 1)$$

where CI and CII are binary encoded variables. In order to analyze a qualitative logical model, one first finds all reachable states using a depth first search assuming some initial state. Assuming that CI and CII are both initially low, the reachable state space is shown in Figure 6.11(a). Namely, CII goes to a high concentration first activating CI production so it can go high. CI represses CII production making it go low which causes CI production to slow and return to a low concentration. It is important to note that this represents only the most likely scenario as it is potentially possible that through basal production of CI, that it goes to a high concentration before CII increases. However, such a coarse analysis can often be useful.

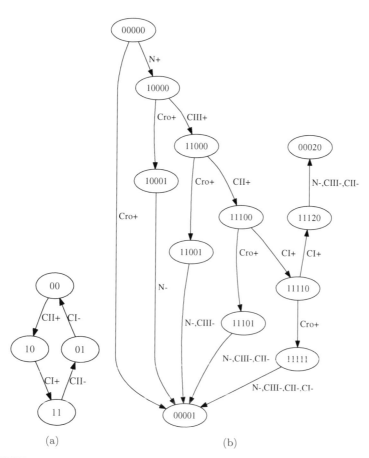

FIGURE 6.11: (a) State graph for the CI/CII portion of the phage λ decision circuit. The label of the state encoding is (CII,CI). (b) Partial state graph for the complete phage λ decision circuit. The label of the state encoding is (N, CIII, CII, CI, Cro).

The full phase λ decision circuit can be represented logically as follows:

$$b_N := (b_{CI} \neq 2) \wedge (b_{Cro} = 0)$$
$$b_{CIII} := (b_N = 1) \wedge (b_{CI} \neq 2) \wedge (b_{Cro} = 0)$$
$$b_{CII} := (b_N = 1) \wedge (b_{CIII} = 1) \wedge (b_{CI} \neq 2) \wedge (b_{Cro} = 0)$$
$$b_{CI} := (((b_{CII} = 1) \wedge (b_{CI} = 0)) + 2(b_{CI} \neq 0)) \wedge (b_{Cro} = 0)$$
$$b_{Cro} = (b_{CI} \neq 2)$$

where N, CII, CIII, and Cro are binary variables, and CI is a ternary variable. Starting from an initial state of all species low, the reachable states are those shown in Figure 6.11(b). In the initial state, both b_{Cro} and b_N are enabled to change from 0 to 1. If b_{Cro} goes to high first, this represses N production meaning that it no longer can go high. In this case, the state (00001) is reached which is the stable lysis state. If, however, b_N goes high first, this enables b_{CIII} to go high. Again, b_{Cro} could go high first which would repress N production and allow degradation to eliminate N. This means that b_N would become enabled to change from 1 to 0 leading again to the stable lysis state (00001). If b_{CIII} goes high first, this allows CIII to protect CII which enables b_{CII} to change from 0 to 1. Again, if b_{Cro} goes high first, N and CIII are repressed and their logical variables become enabled to go low. To simplify the presentation, the concurrent interleavings are omitted from the state graph since they all end up in the same stable lysis state (00001). If b_{CII} goes high first, this enables CI production from P_{RE} meaning that b_{CI} is enabled to change from 0 to 1. Once b_{CI} goes to 1, this enables CI production from P_{RM} which means that b_{CI} is enabled to change from 1 to 2. Once b_{CI} has reached its highest value, N, CII, and CIII are all repressed and degrade back to their low value. This results in the stable lysogeny state (00020).

6.6 Sources

Arkin presents an interesting comparison of electrical and biochemical circuits in Arkin (2000). Several alternative piecewise models have been proposed in Glass (1977); Plahte *et al.* (1994); Gouzé and Sari (2001). The discussion in Sections 6.1 and 6.2 follows the presentation in Gouzé and Sari (2001). One particularly interesting piecewise model is the one proposed for phage λ in McAdams and Shapiro (1995) which inspired Figure 6.7. Matsuno *et al.* proposed the use of HPNs for modeling genetic circuits in Matsuno *et al.* (2000). The LHPN model used in Section 6.2 is from Little *et al.* (2006). The material on stochastic FSMs as well as the phage λ analysis results are from Kuwahara *et al.* (2006); Kuwahara (2007). Stewart's text is an excellent reference on Markov chain analysis, and it motivated the material in Section 6.4 (Stewart, 1994). Boolean logical models of genetic circuits

were proposed in Kauffman (1969); Kauffman *et al.* (1978). Thomas, Thieffry, and others proposed the use of *generalized logical models* (Thomas, 1991;
Thieffry and Thomas, 1995). The main features of these models is that their
behavior is asynchronous and that species may be modeled by n-ary (not
just Boolean) variables. Another interesting qualitative model is *qualitative
differential equations* which are presented in Jong *et al.* (2001).

Problems

6.1   Consider the abstracted reaction-based model from Problem 5.7.

6.1.1. Determine the critical levels for the important species in this model
assuming that the amplifier, a, is 0.95.

6.1.2. Create a PLDE model and find the solution for each logical domain.

6.1.3. Create a stochastic FSM for this genetic circuit.

6.1.4. Analyze the stochastic FSM using Markov chain analysis. In particular, determine its CTMC and EMC representations. Then, using a computer
program, find the steady-state distributions for the EMC and CTMC. Do your
results make sense? Explain your answer.

6.1.5. Construct a qualitative logical model and find its state graph. Do
your results make sense? Explain your answer.

6.2   Consider the abstracted reaction-based model from Problem 5.8.

6.2.1. Determine the critical levels for the important species in this model
assuming that the amplifier, a, is 0.95.

6.2.2. Create a PLDE model and find the solution for each logical domain.

6.2.3. Create a stochastic FSM for this genetic circuit.

6.2.4. Analyze the stochastic FSM using Markov chain analysis. In particular, determine its CTMC and EMC representations. Then, using a computer
program, find the steady-state distributions for the EMC and CTMC. Do your
results make sense? Explain your answer.

6.2.5. Construct a qualitative logical model and find its state graph. Do
your results make sense? Explain your answer.

6.3   Consider the abstracted reaction-based model from Problem 5.9.

6.3.1. Determine the critical levels for the important species in this model
assuming that the amplifier, a, is 0.95.

6.3.2. Create a PLDE model and find the solution for each logical domain.

6.3.3. Create a stochastic FSM for this genetic circuit.

6.3.4. Analyze the stochastic FSM using Markov chain analysis. In particular, determine its CTMC and EMC representations. Then, using a computer
program, find the steady-state distributions for the EMC and CTMC. Do your
results make sense? Explain your answer.

6.3.5. Construct a qualitative logical model and find its state graph. Do
your results make sense? Explain your answer.

6.4 Consider the abstracted reaction-based model from Problem 5.10.

6.4.1. Determine the critical levels for the important species in this model assuming that the amplifier, a, is 0.95.

6.4.2. Create a PLDE model and find the solution for each logical domain.

6.4.3. Create a stochastic FSM for this genetic circuit.

6.4.4. Analyze the stochastic FSM using Markov chain analysis. In particular, determine its CTMC and EMC representations. Then, using a computer program, find the steady-state distributions for the EMC and CTMC. Do your results make sense? Explain your answer.

6.4.5. Construct a qualitative logical model and find its state graph. Do your results make sense? Explain your answer.

6.5 Consider the abstracted reaction-based model from Problem 5.11.

6.5.1. Determine the critical levels for the important species in this model assuming that the amplifier, a, is 0.95.

6.5.2. Create a PLDE model and find the solution for each logical domain.

6.5.3. Create a stochastic FSM for this genetic circuit.

6.5.4. Analyze the stochastic FSM using Markov chain analysis. In particular, determine its CTMC and EMC representations. Then, using a computer program, find the steady-state distributions for the EMC and CTMC. Do your results make sense? Explain your answer.

6.5.5. Construct a qualitative logical model and find its state graph. Do your results make sense? Explain your answer.

6.6 Use iBioSim to perform logical abstraction and analysis of the model from Problem 1.8.

6.7 Use iBioSim to perform logical abstraction and analysis of the model from Problem 1.9.

Chapter 7

Genetic Circuit Design

Design is not just what it looks like and feels like. Design is how it works.

—Steve Jobs

Design is so simple, that's why it is so complicated.

—Paul Rand

One of the goals of this textbook is to describe how electrical engineering methods can be applied to better understand the behavior of genetic circuits. If this objective is achievable, then it may also be possible to design *synthetic genetic circuits* that behave like particular electrical circuits such as switches, *oscillators*, and sequential finite state machines. Synthetic genetic circuits have the potential to help us better understand how microorganisms function, produce drugs more economically, metabolize toxic chemicals, and even modify bacteria to hunt and kill tumors. *Genetic engineering* has been around for more than 30 years, and it is based on the following three techniques:

- *Recombinant DNA* - constructing artificial DNA through combinations.

- *Polymerase chain reaction* (PCR) - making many copies of this DNA.

- *Automated sequencing* - checking the resulting DNA sequence.

Recently, a new field, synthetic biology has added the following ideas:

- *Automated construction* - separate design from construction.

- *Standards* - create repositories of parts that can be easily composed.

- *Abstraction* - high-level models to facilitate design.

This chapter first presents in Section 7.1 experimental techniques to assemble synthetic genetic circuits. Next, this chapter presents several small synthetic genetic circuit examples. In particular, Section 7.2 presents designs for several combinational logic gates such as AND gates and OR gates. Section 7.3 presents an alternative method of constructing combinational logic gates known as *PoPS gates*. Section 7.4 presents designs of sequential, state-holding logic gates including a *genetic Muller C-element*. This chapter concludes with Section 7.5 which presents some of the remaining challenges for synthetic biology.

7.1   Assembly of Genetic Circuits

A registry of standard biological parts is being developed at:

http://partsregistry.org

This registry includes descriptions of numerous $BioBricks^{TM}$ as shown in Figure 7.1(a). $BioBrick^{TM}$ part types include: *terminators, ribosome binding sites*, protein coding regions (i.e., genes), reporter genes, signaling parts, regulatory sequences, gates, etc. $BioBricks^{TM}$ can be assembled into more complex devices and systems as shown in Figure 7.1(b). The $BioBrick^{TM}$ assembly process is depicted in Figure 7.1(c). Each $BioBrick^{TM}$ has an E and X domain before the $BioBrick^{TM}$ and an S and P domain after the $BioBrick^{TM}$. To combine two $BioBricks^{TM}$, the first $BioBrick^{TM}$ is cut with proteases that cut the E and S domains while the second one is cut at the E and X domains. The results are then mixed and ligated resulting in the exposed E domains binding and the S and X domains binding resulting in the final combined $BioBrick^{TM}$.

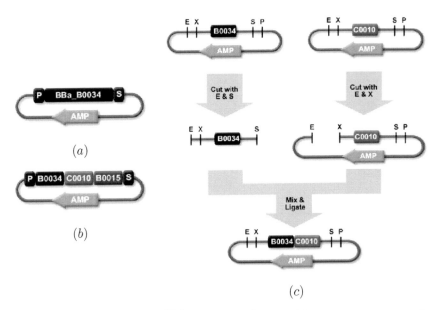

FIGURE 7.1: $BioBricks^{TM}$ (courtesy of http://partsregistry.org). (a) An example $BioBrick^{TM}$. (b) A more complex device composed of three $BioBricks^{TM}$. (c) $BioBrick^{TM}$ assembly process.

7.2   Combinational Logic Gates

Combinational logic gates are ones in which their output is strictly determined by their current input. In other words, these circuits do not include memory and typically do not include feedback. The simplest combinational gate is an *inverter* which can be constructed from genetic material as shown in Figure 7.2. In this circuit, the input is the TetR protein, and the output is green fluorescent protein (GFP) shown schematically in Figure 7.2(a). Cells that include this inverter glow green when TetR is not present within the cell. This behavior can be expressed using a truth table such as the one shown in Figure 7.2(b) in which a 0 indicates that the protein is not present in a significant concentration, and a 1 indicates that it is present in a significant concentration. Figure 7.2(c) shows the genetic implementation of this inverter. This gate is composed of a promoter that is repressed by TetR and the *GFP* gene. Figure 7.2(d) shows the average of 500 stochastic simulation runs for this gate which shows that this genetic gate behaves like an inverter.

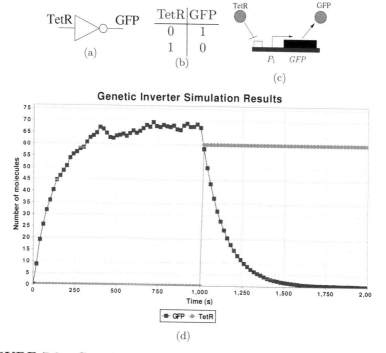

FIGURE 7.2: Genetic inverter. (a) Schematic symbol. (b) Truth table. (c) Genetic implementation. (d) Stochastic simulation.

If an odd number of inverters are connected in a ring as shown in Figure 7.3(a), they form an oscillator. Oscillations are often used as central clocks to synchronize behavior. A biological example is *circadian rhythms* which manifest as periodic variations of concentrations of particular proteins within a cell. Though the precise mechanism that causes these rhythms is unknown, Elowitz and Leibler constructed the synthetic genetic circuit shown in Figure 7.3(b) known as the *repressilator* that has a similar behavior. In this circuit, the choices of parameter values are very important to produce oscillations. For example, high protein synthesis and degradation rates, large cooperative binding effects, and efficient repression are all necessary. Therefore, strong and tightly repressible promoters are selected, and proteins are modified to make them easy targets for proteases. The average of 500 stochastic simulation runs is shown in Figure 7.3(c). These results show that the fluorescence intensity is oscillatory.

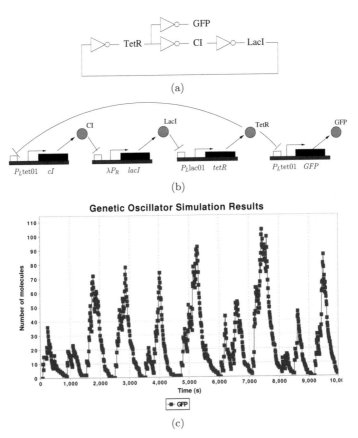

(a)

(b)

Genetic Oscillator Simulation Results

(c)

FIGURE 7.3: Genetic oscillator. (a) Logic diagram.
(b) Genetic implementation. (c) Simulation results.

Another important combinational gate is the 2-input *NAND gate*. It is known as a *universal gate*, since given a collection of them, it is possible to construct any other combinational gate. The symbol for a NAND gate is shown in Figure 7.4(a) with input proteins LacI and TetR and output GFP, and its behavior is defined by the truth table in Figure 7.4(b). Namely, GFP is produced as long as either LacI or TetR is not present in the cell. A genetic NAND gate can be constructed using two inverters with the same gene as shown in Figure 7.4(c). If either LacI or TetR is not present, then at least one *GFP* gene actively produces GFP. Stochastic simulation shown in Figure 7.4(d) indicates that it behaves like a NAND gate. A 2-input *NOR gate* is another universal gate, and a genetic implementation is shown in Figure 7.5 which uses a promoter that can be repressed by either LacI or TetR.

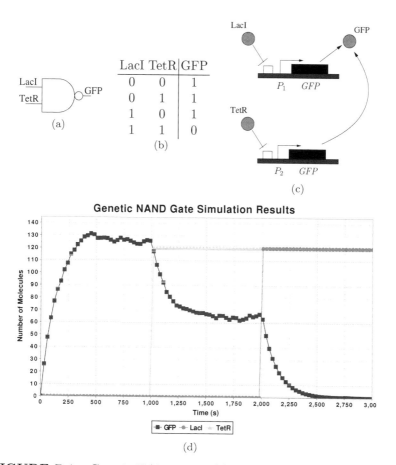

FIGURE 7.4: Genetic NAND gate. (a) Schematic symbol. (b) Truth table. (c) Genetic implementation. (d) Stochastic simulation.

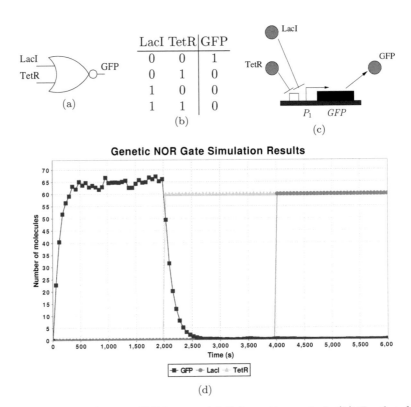

FIGURE 7.5: Genetic NOR gate. (a) Schematic symbol. (b) Truth table.
(c) Genetic implementation. (d) Stochastic simulation.

As mentioned above, it is possible to construct more complicated combinational logic gates from simpler ones. For example, Figure 7.6 shows the design of an *AND gate* that is constructed from a NAND gate followed by an inverter. In this gate, unless both LacI and TetR are present then CI is produced from at least one gene repressing the production of GFP. An *OR gate* can be constructed from two inverters and a NAND gate as shown in Figure 7.7. In this case, if LacI or TetR is present then GFP is produced. Namely, LacI represses production of CI allowing GFP to be produced from the top *GFP* gene while TetR represses production of Cro allowing GFP to be produced from the bottom *GFP* gene.

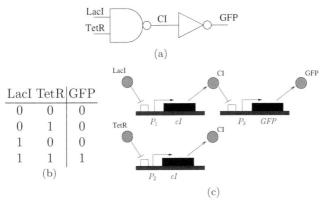

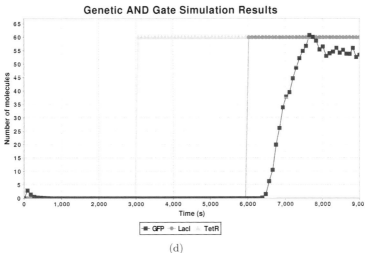

FIGURE 7.6: Genetic AND gate using a NAND gate and inverter. (a) Logic diagram. (b) Truth table. (c) Genetic implementation. (d) Stochastic simulation.

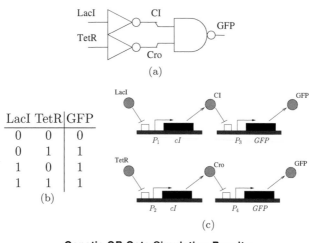

(a)

LacI	TetR	GFP
0	0	0
0	1	1
1	0	1
1	1	1

(b)

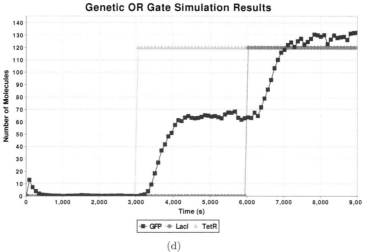

(c)

(d)

FIGURE 7.7: Genetic OR gate using two inverters and a NAND gate. (a) Logic diagram. (b) Truth table. (c) Genetic implementation. (d) Stochastic simulation.

Figure 7.8 shows a genetic AND gate that uses as inputs two small molecule chemical inducers, *isopropyl β-D-thiogalatopyranoside* (IPTG) and *anhydrotetracycline* (aTc). IPTG disables LacI's ability to act as a repressor by preventing it from being able to bind to the promoter. The inducer aTc has the same effect on TetR. In this circuit, P_1 is always active producing the proteins LacI and TetR, and P_2 can be repressed by either LacI or TetR. If, however, both IPTG and aTc are added then neither can bind to P_2, and GFP is produced. A genetic OR gate using chemical inducers is shown in Figure 7.9 in which P_2 is repressed by LacI while P_3 is repressed by TetR. Therefore, GFP is produced from P_2 when IPTG is present and from P_3 when aTc is present.

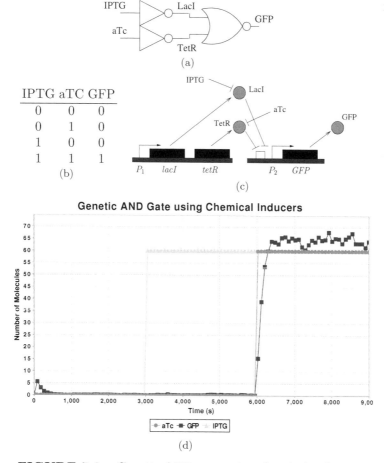

IPTG	aTC	GFP
0	0	0
0	1	0
1	0	0
1	1	1

(b)

FIGURE 7.8: Genetic AND gate using chemical inducers.
(a) Logic diagram. (b) Truth table. (c) Genetic implementation.
(d) Stochastic simulation.

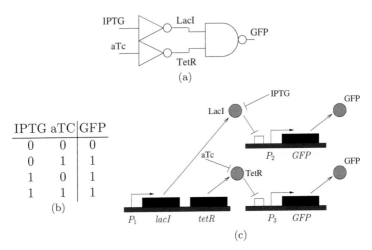

IPTG	aTC	GFP
0	0	0
0	1	1
1	0	1
1	1	1

(b)

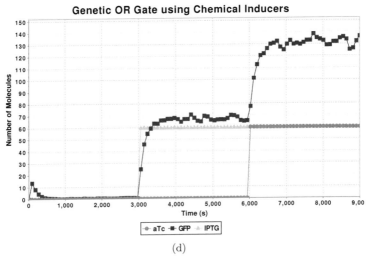

(d)

FIGURE 7.9: Genetic OR gate using chemical inducers.
(a) Logic diagram. (b) Truth table. (c) Genetic implementation.
(d) Stochastic simulation.

All of the circuits just described require several promoters and genes making them quite complex. An AND gate implementation that requires only a single promoter and gene is shown in Figure 7.10. The inputs to this gate are the proteins LuxR and LuxI. The protein LuxI can be converted into $3OC_6HSL$ which can then bind with LuxR to produce a complex. This complex can activate the promoter shown in Figure 7.10(c) to activate transcription. In other words, as shown in the stochastic simulation in Figure 7.10(d), both LuxI and LuxR are necessary for significant production of GFP.

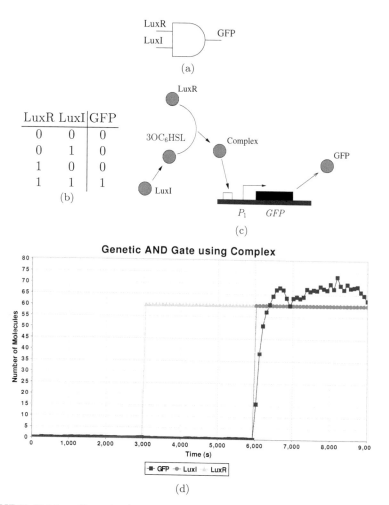

FIGURE 7.10: Genetic AND gate using one gene. (a) Schematic symbol. (b) Truth table. (c) Genetic implementation. (d) Stochastic simulation.

7.3 PoPS Gates

Connecting gates can be difficult because each gate uses different proteins as inputs and outputs. One possible solution is to use the rate of gene expression as the signal much like electrical current which is the rate of electron flow. The rate of gene expression is the number of RNA polymerases that move across a DNA strand per second (i.e., PoPS). A PoPS inverter is shown in Figure 7.11(a). A PoPS inverter has the gene and promoter reversed. In particular, when there are PoPS input, the repressor gene is transcribed. This produces a repressor molecule that represses the operator which stops PoPS from being output. Similarly, if there is no PoPS input, then the repressor is not produced allowing for RNAP to bind to the operator leading to PoPS output. A NOR gate can be constructed by having two repressor genes that receive PoPS input from two different input sources as shown in Figure 7.11(b). This gate only produces PoPS output when there is no PoPS on either input. An OR gate can be constructed by changing the genes to produce an activator instead of a repressor as shown in Figure 7.11(c). A PoPS NAND gate can be constructed by using two genes that each produce half of a repressor molecule which only repress the operator when both halves are produced and bind together as shown in Figure 7.11(d). A PoPS AND gate can be constructed by changing the repressor genes to activator genes as shown in Figure 7.11(e). The last PoPS example shown in Figure 7.11(f) show PoPS gates that can be used to communicate a signal from one cell to another. The genetic sender constructs a signaling molecule when there is PoPS input. The genetic receiver produces PoPS output when there is PoPS input and the signaling molecule is present to bind to an activator molecule.

7.4 Sequential Logic Circuits

The output of sequential logic circuits depend not only on the current input, but also on the recent history of inputs. This history is recorded in the state of the circuit. This state is maintained through the use of *feedback*. Feedback loops are important for achieving stability in control systems. In *autoregulation*, a protein modifies its own rate of production. Feedback can be either positive or negative. *Negative feedback* can enhance stability. *Positive feedback*, on the other hand, can be used to create *bistability*. In phage λ, for example, the protein CI enhances its own production from the promoter P_{RM} which helps create the two stable states of lysis and lysogeny.

Figure 7.12 shows how feedback can change a combinational circuit into a sequential circuit. In this circuit, the AND gate from Figure 7.8 is modified

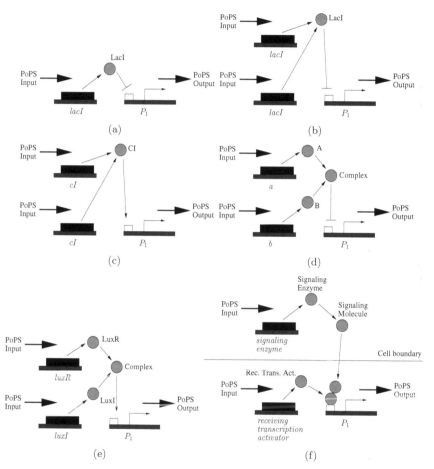

FIGURE 7.11:   PoPS examples: (a) inverter, (b) NOR gate, (c) OR gate, (d) NAND gate, (e) PoPS AND gate, and (f) PoPS cell-to-cell communication circuit.

by adding the *cI* gene to be transcribed with the *GFP* gene from P_2. Also, P_1 is modified to be repressed by the CI protein. The result is that this AND gate now has memory. If at anytime IPTG and aTc are both present, then CI is produced which represses P_1. This means even after IPTG and aTc are removed, the cell continues to glow, since LacI and TetR cannot be produced allowing for P_2 to remain active.

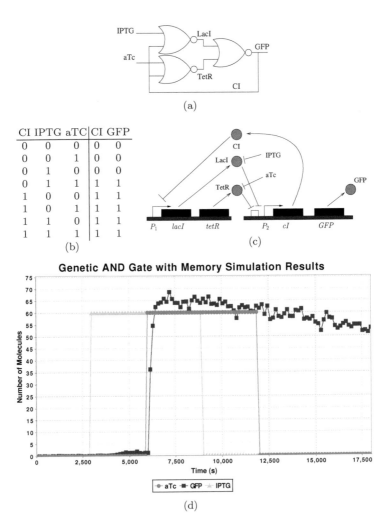

CI	IPTG	aTC	CI	GFP
0	0	0	0	0
0	0	1	0	0
0	1	0	0	0
0	1	1	1	1
1	0	0	1	1
1	0	1	1	1
1	1	0	1	1
1	1	1	1	1

(b)

(c)

(d)

FIGURE 7.12: Genetic AND gate with memory. (a) Logic diagram. (b) Truth table. (c) Genetic implementation. (d) Stochastic simulation.

In the previous circuit, once the GFP level is high, it remains high forever. A more useful circuit is one in which the state can be reset back to the low level. A circuit with this behavior is the genetic *toggle switch* shown in Figure 7.13. A toggle switch has a set input, S, a reset input, R, and a state output, Q. The circuit shown in Figure 7.13(a) uses IPTG for S, heat for R, and GFP for Q. The behavior of this circuit is shown in the truth table in Figure 7.13(b). Namely, if heat is applied, then the GFP level goes low. If IPTG is applied, the GFP level goes high. If neither heat or IPTG are applied, then the GFP level retains its previous state. Note that if both IPTG and heat are applied, the final GFP state is indeterminate. The genetic toggle switch implementation is shown in Figure 7.13(c), and stochastic simulation results are shown in Figure 7.13(d). With an initially high level of LacI, the GFP level is low. After IPTG is provided, LacI is no longer repressing the promoter P_{trc-2} and CITS (temperature sensitive CI) and GFP are produced. CITS now represses promoter P_{Ls1con} so even after IPTG is removed, GFP expression remains high. When heat is applied, CITS no longer represses P_{Ls1con} allowing LacI to buildup and repress P_{trc-2} resulting in a reduction in the CITS and GFP levels. Again, after heat is removed, this state is maintained.

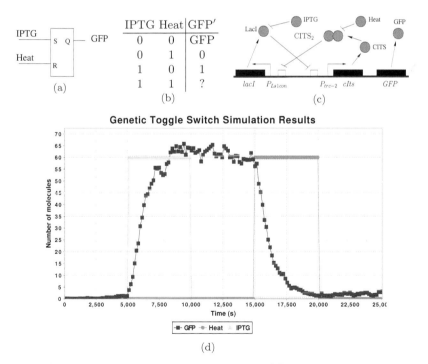

(a)

IPTG	Heat	GFP'
0	0	GFP
0	1	0
1	0	1
1	1	?

(b)

(c)

(d)

FIGURE 7.13:   Genetic toggle switch. (a) Schematic symbol. (b) Truth table. (c) Genetic implementation. (d) Stochastic simulation.

Another useful sequential logic gate is the *Muller C-element*. The Muller C-element is a state-holding gate that is commonly used in many asynchronous digital design methods. As shown in the truth table in Figure 7.14(a), the behavior of a Muller C-element is when both inputs are low, the output is low, when both inputs are high, the output is high, and otherwise it retains its previous state. A Muller C-element can be utilized to synchronize multiple independent processes. The design of a genetic Muller C-element can potentially enable the design of any asynchronous finite state machine.

Combining the toggle switch with combinational logic as shown in Figure 7.14(b) produces the desired behavior of a Muller C-element. Its genetic implementation is shown in Figure 7.14(c). The average result of 1000 stochastic simulation runs is shown in Figure 7.14(d) which show that the average behavior is indeed that of a C-element. In other words, the output matches the inputs when they agree and retains its previous state when they are mixed.

Another way of looking at this design is using nullcline analysis as shown in Figure 7.15. These nullclines consider the species Y and Z which indicate the state of the toggle switch. When [Y] is high, [C] is also high, since Y and C are generated from the same promoter. Similarly, when [Z] is high, [C] is low, since Z represses C production. Figure 7.15(a) shows the nullclines (i.e., the lines indicating where the rates of change of [Y] and [Z] are zero) when the concentrations of the inputs, A and B, are low. These nullclines intersect to create a stable equilibrium point with [Z] at a high level and [Y] at a low level (i.e., the output C is low). Figure 7.15(b) shows the nullclines when the concentrations of the inputs, A and B, are high, and in this case there is a stable equilibrium point with [Z] at a low level and [Y] at a high level (i.e., the output C is high). Finally, Figure 7.15(c) shows the nullclines when the inputs, A and B, are mixed. In this case, there are two stable equilibrium points and one unstable equilibrium point. Which stable state is reached depends on the state before the inputs become mixed. In other words, if the inputs are low before becoming mixed, the stable state reached is the one in which [Z] is high while if the inputs are previously both high, the stable state reached is the one in which [Y] is high.

While gates constructed from silicon technology tend to have an extremely low probability of failure, gates constructed from genetic technology have a significant probability of failure. A state-holding gate is most susceptible to errors when there are multiple stable states. For example, when both inputs are low or both are high, it is possible that stochastic effects move the gate away from the stable equilibrium point, but the gate eventually is driven back towards this state. On the other hand, when the inputs are mixed, stochastic effects could cause the circuit to move erroneously from one stable state into the other stable state. Figure 7.16(a) shows the likelihood of such a failure for 1000 stochastic runs over one cell cycle (i.e., 2100 seconds). This plot compares two definitions of failure. In the *single-rail* definition, a failure is defined when the concentration of the output C goes low when it should stay high or vice versa. In the *dual-rail* definition, a failure is defined as when both

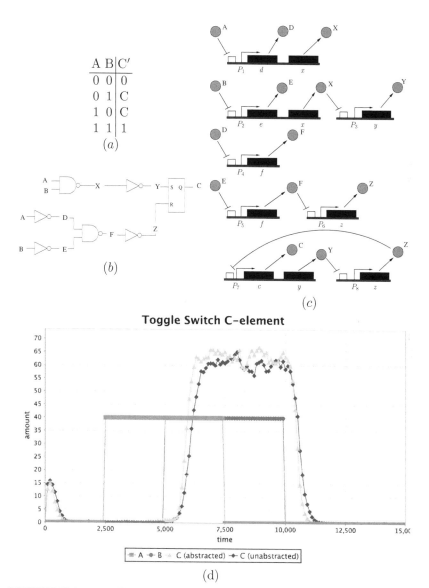

FIGURE 7.14: Genetic toggle Muller C-element (courtesy of Nguyen (2008)). (a) Truth table. (b) Logic diagram. (c) Genetic implementation. (d) Stochastic simulation.

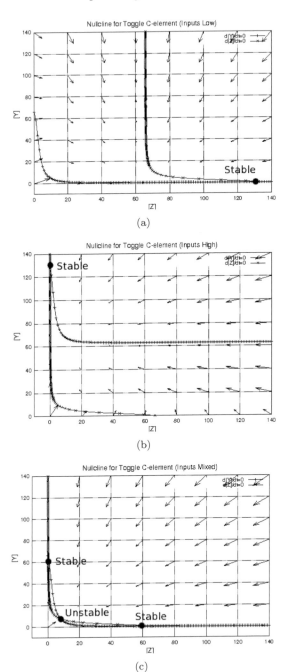

FIGURE 7.15:   Nullcline analysis for the genetic toggle C-element
(courtesy of Nguyen (2008)). (a) Both inputs low, (b) both inputs high, and
(c) inputs are mixed.

the concentrations of X and Y change values. Under the single-rail definition, the toggle genetic C-element fails 15 percent of the time while under dual-rail, the failure rate is only about 0.5 percent. This difference occurs because [C] may drop temporarily due to stochastic effects, but it may recover since the state is still closer to the correct stable state.

There are several different alternative ways to design a C-element. One of the simplest is to use what is known as a *majority gate*. A majority gate is a gate that outputs high when the majority of its inputs are high. A 3-input majority gate configured with 2-inputs as the primary inputs A and B and a third input as a feedback state signal results in a C-element. Such a gate can be constructed from NAND gates and inverters as shown in Figure 7.17(a). Note that the inverters are not strictly needed logically, but they are added so that the feedback signal is not used to directly drive other gates. With some optimizations, this gate can be implemented as a genetic circuit as shown in Figure 7.17(b), and the average of 1000 stochastic runs shown in Figure 7.17(c) indicate that this gate has the desired behavior. Another alternative logic design of a C-element is shown in Figure 7.18(a). This gate is known as a *speed-independent* Muller C-element because this gate operates correctly regardless of the delay of any gate in the design. The genetic circuit implementation for the speed-independent Muller C-element is shown in Figure 7.18(b), and the average results for 1000 stochastic runs are shown in Figure 7.18(c).

Given these three alternative designs which design is the best? To determine this, let us consider the failure rates of each of these designs as shown in Figure 7.16(b). These results show that the worst design is the majority C-element which is not surprising as it has tight timing constraints that it must meet for correctness. The perhaps surprising result is that the toggle C-element is more robust than the speed-independent design. This result indicates that timing robustness does not necessarily provide robustness to stochastic effects as extra circuitry added for timing robustness may actually make it more susceptible to stochastic effects.

The next question to ask is how do these gates perform as parameters vary. The first parameter considered is the degree of cooperativity (i.e., how many molecules of each transcription factor must bind to repress or activate a promoter). The results shown in Figure 7.19(a) indicate that a cooperativity of at least two is necessary for reliable operation. This result is due to the fact that a cooperativity of one yields linear rate behavior while two results in non-linear behavior which is essential to state-holding. Figure 7.19(b) shows that higher repression strength (i.e., how tightly a transcription factor binds) yields more robust behavior. Figure 7.20(a) shows that higher decay rates yield less robustness. In this case, care must be taken as Figure 7.20(b) shows that *switching time* (i.e., how fast the state of the C-element can be changed) increases as decay rate is decreased. In other words, one must select decay rates that are low enough to achieve reasonable robustness but not so low that the gate switches state too slowly.

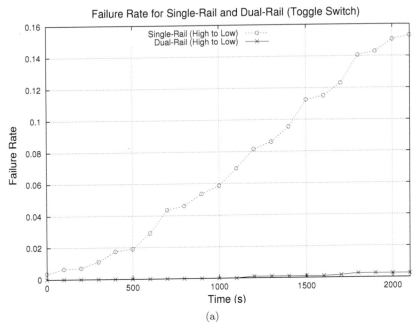

(a)

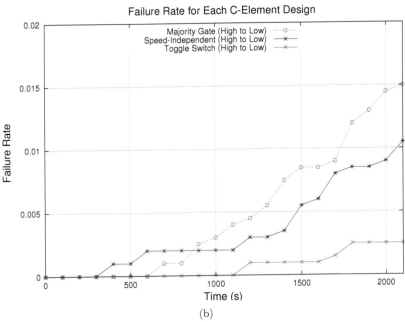

(b)

FIGURE 7.16: Stochastic analysis for the genetic C-elements (courtesy of Nguyen (2008)). (a) Failure rate for single-rail and dual-rail interpretation of the gate output. (b) Failure rate for each genetic C-element design.

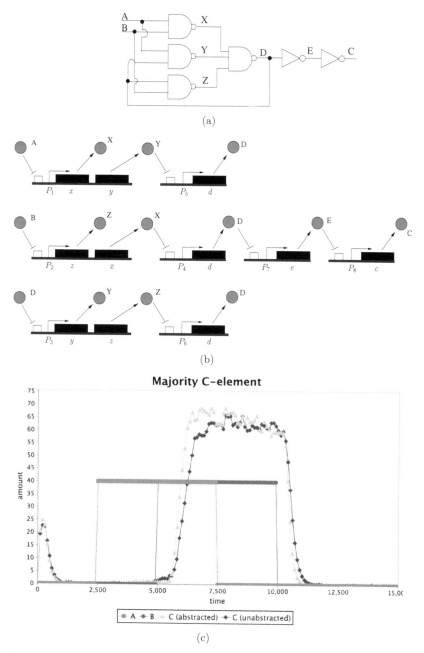

(a)

(b)

Majority C-element

(c)

FIGURE 7.17: Genetic majority Muller C-element (courtesy of Nguyen
(2008)). (a) Logic diagram. (b) Genetic implementation.
(c) Stochastic simulation.

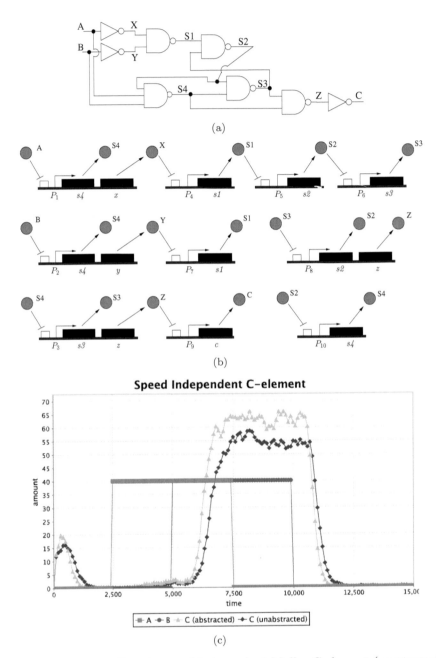

FIGURE 7.18: Genetic speed-independent Muller C-element (courtesy of Nguyen (2008)). (a) Logic diagram. (b) Genetic implementation. (c) Stochastic simulation.

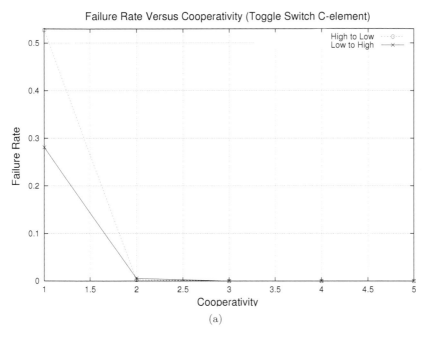

(a)

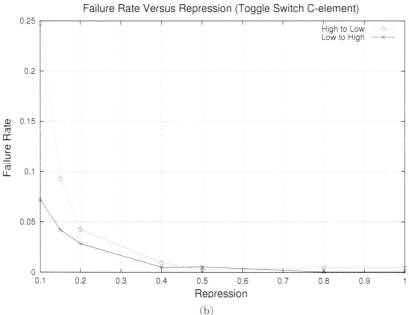

(b)

FIGURE 7.19: Effects of (a) cooperativity and (b) repression strength on failure rate of the genetic toggle C-element (courtesy of Nguyen (2008)).

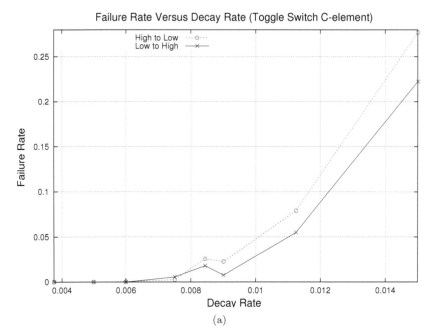

(a)

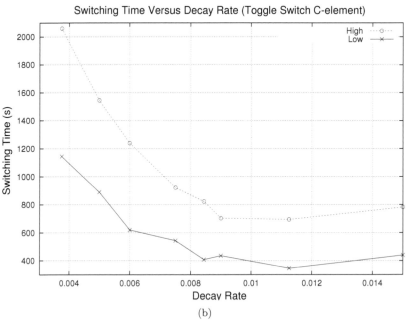

(b)

FIGURE 7.20:   Effects of decay rate on the toggle switch genetic
C-element. (courtesy of Nguyen (2008)). (a) Failure rate for various decay
rates. (b) Switching time for various decay rates.

7.5 Future Challenges

An increasing number of laboratories are designing more ambitious and mission critical synthetic biology projects. Synthetic biology has the potential to help us better understand how microorganisms function by allowing us to examine differences in how a synthetic pathway behaves *in vivo* as compared with *in silico* (Sprinzak and Elowitz, 2005). There are also numerous exciting potential applications. The Gates Foundation funds research on the design of pathways for production of antimalarial drugs (Ro *et al.*, 2006). Scientists have also been working on modifying bacteria to metabolize toxic chemicals (Brazil *et al.*, 1995; Cases and de Lorenzo, 2005). Finally, a number of labs are designing bacteria to hunt and kill tumors (Anderson *et al.*, 2006).

While these projects represent tremendous progress and exciting possibilities, there are still numerous challenges to be faced before we can truly engineer genetic circuits. First, there are no wires in genetic circuits meaning that they currently have no signal isolation. Without signal isolation, proteins produced by the gates may interfere with each other as well as the normal operation of the host cell. This fact also has the side-effect that gates in a genetic circuit library can typically only be used once, since two copies of the same gate would use the same proteins and interfere with each other. Second, the behavior and timing of genetic circuits is non-deterministic and difficult to characterize. The probability of failure of genetic gates likely can never be reduced to that of silicon gates. As described in the previous section, tuning of cooperativity, repression strength, and decay rates can improve robustness, but unfortunately the control of these parameters is likely to be quite limited. This fact means that genetic circuits will be designed using unreliable components that have wide variations in parameter values. These circuits will also likely utilize asynchronous timing, since it is unlikely that a global clock to synchronize events makes sense in a genetic technology.

Design using silicon technology also faced similar challenges in its early days, and it continues to face new challenges every year. While some challenges are addressed through improvements in manufacturing processes, many of them are solved through the development of sophisticated *electronic design automation* (EDA) tools. These EDA tools manage design complexity so that ever more complicated integrated circuits can be designed each year. It is our belief that crucial to the success of synthetic biology is an improvement in methods and tools for *genetic design automation* (GDA). One of the major goals of this textbook is to help foster the development of such GDA tools.

Most GDA tools currently require biologists to model, analyze, and design genetic circuits at the molecular level using languages such as SBML. This representation is at a very low level which is roughly equivalent to the layout level for electronic circuits. Designing and simulating genetic circuits at this level of detail is extremely tedious and time-consuming. Therefore, another major goal of this textbook has been to introduce higher-level abstractions that can be used for modeling, analysis, and design of genetic circuits.

Finally, as mentioned above, genetic circuits utilize asynchronous timing. Some of the analysis methods presented in this book have been inspired by our previous research in asynchronous circuit design (Myers, 2001). In the future, we believe that synthesis methods for asynchronous circuit design will also likely be successfully adapted to the requirements for design using a genetic circuit technology.

Interestingly, future silicon and proposed nano-devices will likely face a number of the same challenges that we currently face with a genetic technology. Namely, their devices are becoming more non-deterministic in behavior and timing, and they are thus less reliable. Therefore, for Moore's law to continue, future design methods will need to support the design of reliable systems using unreliable components. This fact means that GDA tools may potentially be useful to produce future integrated circuit technologies. This may have the added benefit that these new circuits may not only be more robust but also more power efficient. For example, the human inner ear performs the equivalent of one billion floating point operations per second and consumes only 14 μW while a game console with similar performance burns about 50 W (Sarpeshkar, 2006). It is our belief that this difference of six orders of magnitude in power efficiency is due to over designing components in order to achieve the necessary extremely low probability of failure in every device. These facts lead to the following observation:

> Since the engineering principles by which such circuitry is constructed in cells comprise a super-set of that used in electrical engineering, it is, in turn, possible that we will learn more about how to design asynchronous, robust electronic circuitry as well (Arkin, 2000).

7.6   Sources

Some recent interesting reviews that describe the field of synthetic biology appear in Endy (2005); Canton *et al.* (2008); Arkin (2008). These reviews describe the methods being developed to standardize and assemble BioBricks^{TM} as well as the PoPS gate design methodology. A good review of genetic logic design can be found in Hasty *et al.* (2002). An interesting experimental study of genetic logic gates appears in Guet *et al.* (2002). The genetic oscillator known as the repressilator is described in Elowitz and Leibler (2000). The genetic toggle is described in Gardner *et al.* (2000). Finally, the genetic Muller C-element design and its analysis appears in Nguyen *et al.* (2007); Nguyen (2008).

Problems

7.1   An XOR gate is one in which its output goes high when exactly one of its inputs is high. The behavior of an XOR gate is defined by the following truth table:

A	B	C
0	0	0
0	1	1
1	0	1
1	1	0

7.1.1. Design and simulate a genetic XOR gate.

7.1.2. Design a PoPS XOR gate.

7.2   An AOI21 (or AND-OR-INVERT21) gate is one in which its output goes low either when both A and B are high or C is high. The behavior of an AOI21 gate is defined by the following truth table:

A	B	C	D
0	0	0	1
0	0	1	0
0	1	0	1
0	1	1	0
1	0	0	1
1	0	1	0
1	1	0	0
1	1	1	0

7.2.1. Design and simulate a genetic AOI21 gate.

7.2.2. Design a PoPS AOI21 gate.

7.3   Design and simulate a sequential circuit with one input and one output that toggles its output every time it receives a pulse on its input. In other words, the output goes high after the input goes high, and it stays high until the input goes back to low and then to high again.

7.4   Design and simulate a D-type latch. This is a sequential circuit with two inputs, D and Clk, and one output, Q. When Clk is high, the output Q should have the same state as D. When Clk is low, the output Q should not change.

Solutions to Selected Problems

1.1 Consider the first part of the enzymatic reaction:

$$E + S \underset{k_2}{\overset{k_1}{\rightleftarrows}} ES$$

Assume that k_1 is $0.01\text{sec}^{-1}nM^{-1}$, k_{-1} is 0.1sec^{-1}, $[E]$ is 35 nM, $[S]$ is 100nM, $[ES]$ is 50nM, and $RT = 0.5961$ kcal mol^{-1} (i.e., $T = 300°$K). Determine the change in the Gibb's Free Energy for the forward reaction. Is the forward or reverse reaction favored? Using trial-and-error find the point (the concentrations of $[E]$, $[S]$, and $[ES]$) in which this reaction reaches a steady-state (hint: remember that for every nM that you add to $[ES]$, you must take an equal amount off of $[E]$ and $[S]$).

Solution:

$$\Delta G = RT \ln \frac{k_2[ES_1]}{k_1[E][S]}$$

$$\Delta G = 0.5961 \ln \frac{0.1(50)}{0.01(35)(100)}$$

$$\Delta G = -1.16 \text{ kcal mol}^{-1}$$

Negative so the forward reaction is favored. $[ES] = [S] = 75$ and $[E] = 10$ results in $\Delta G = 0$.

1.3 Construct a reaction-based SBML model for the genetic circuit shown below using iBioSim or any other tool that includes an SBML editor. Use the parameter values provided and assume that CI dimerizes before acting as a transcription factor.

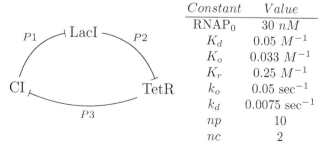

Constant	Value
RNAP$_0$	30 nM
K_d	0.05 M^{-1}
K_o	0.033 M^{-1}
K_r	0.25 M^{-1}
k_o	0.05 sec^{-1}
k_d	0.0075 sec^{-1}
np	10
nc	2

Solution: See Figure 8.1.

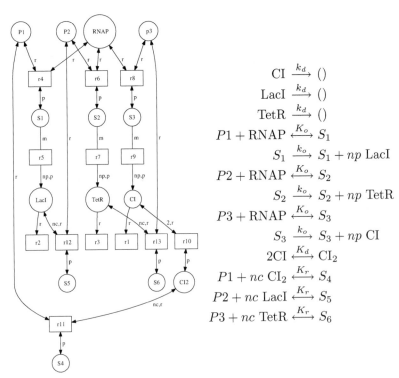

$$CI \xrightarrow{k_d} ()$$
$$LacI \xrightarrow{k_d} ()$$
$$TetR \xrightarrow{k_d} ()$$
$$P1 + RNAP \xleftrightarrow{K_o} S_1$$
$$S_1 \xrightarrow{k_o} S_1 + np \, LacI$$
$$P2 + RNAP \xleftrightarrow{K_o} S_2$$
$$S_2 \xrightarrow{k_o} S_2 + np \, TetR$$
$$P3 + RNAP \xleftrightarrow{K_o} S_3$$
$$S_3 \xrightarrow{k_o} S_3 + np \, CI$$
$$2CI \xleftrightarrow{K_d} CI_2$$
$$P1 + nc \, CI_2 \xleftrightarrow{K_r} S_4$$
$$P2 + nc \, LacI \xleftrightarrow{K_r} S_5$$
$$P3 + nc \, TetR \xleftrightarrow{K_r} S_6$$

FIGURE 8.1: Chemical reaction-based model for Problem 1.1.

2.3 The following questions use the data in this table:

lacI	cI	tetR	gfp	Probability	BN
0	0	0	0	0.04	0.04
0	0	0	1	0.13	0.08
0	0	1	0	0.04	0.06
0	0	1	1	0.08	0.12
0	1	0	0	0.02	0.08
0	1	0	1	0.02	0.02
0	1	1	0	0.18	0.11
0	1	1	1	0.01	0.02
1	0	0	0	0.08	0.07
1	0	0	1	0.16	0.15
1	0	1	0	0.03	0.02
1	0	1	1	0.03	0.05
1	1	0	0	0.09	0.11
1	1	0	1	0.03	0.02
1	1	1	0	0.05	0.04
1	1	1	1	0.01	0.01

2.3.1. Based on the data above, select a Bayesian network.
Solution: There are many possible solutions. One possible Bayesian network is shown in Figure 8.2.

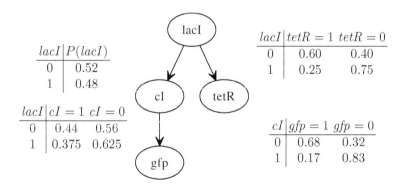

FIGURE 8.2: A Bayesian network model for Problem 2.3.

2.3.2. Annotate each node with its corresponding conditional distribution.
Solution: The conditional distribution for the network selected in the solution for Problem 2.3.1 is shown in Figure 8.2.
2.3.3. Provide the equation for $P(lacI, cI, tetR, gfp)$ implied by the network.
Solution: $P(lacI)P(cI \mid lacI)P(tetR \mid lacI)P(gfp \mid cI)$.

2.3.4. Calculate the probability of each state using your Bayesian network. How well does this compare with the original probabilities?

Solution: The solution is shown in the column labeled BN in the table above. The results compare pretty well except in states 0001, 0011, 0100, and 0110.

2.3.5. Using this information, guess the causal network.

Solution: From the information for the guessed Bayesian network, one would guess that *lacI* represses *cI* and *tetR*, since the likelihood that *cI* or *tetR* is high is less when *lacI* is high as compared to when it is low. One would also guess that *cI* represses *gfp*, since the likelihood that *gfp* is high is considerably less when *cI* is high as compared to when *cI* is low. The actual network that generated this data is shown in Figure 8.3.

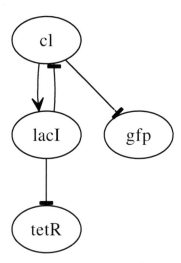

FIGURE 8.3: Causal network for Problem 2.3.5.

2.7 Generate synthetic time series data for the genetic circuit from Problem 1.3 using a stochastic simulator such as the one in `iBioSim`. Collect data for 20 runs over 22 equally spaced time points from 0 to 2100 seconds. Use the causal learning method in `iBioSim` to learn a genetic circuit model.

Solution: If done correctly, this should result in the exact genetic circuit model from Problem 1.3. Namely, the generate model should show that *cI* represses *lacI*, *lacI* represses *tetR*, and *tetR* represses *cI*.

3.1   Consider the following reactions:

$$2S_1 \xrightarrow{0.1} 2S_2$$
$$S_1 + S_2 \xrightarrow{0.2} 2S_1$$

3.1.1.   Determine the reaction rate equations for $[S_1]$ and $[S_2]$.
Solution:

$$\frac{d[S_1]}{dt} = 0.2[S_1][S_2] - 0.2[S_1]^2$$
$$\frac{d[S_2]}{dt} = 0.2[S_1]^2 - 0.2[S_1][S_2]$$

3.1.2.   Simulate by hand using Euler's method the following set of differential equations for 1 second with a time step of 0.2 seconds starting with initial concentrations of $[S_1] = 3.0$ and $[S_2] = 5.0$.
Solution: The simulation result is shown below. Note that the new values of $[S_1]$ and $[S_2]$ are determined by taking the old value and adding to it the present rate of change times the time step of 0.2 seconds.

Time	$[S_1]$	$[S_2]$	$d[S_1]/dt$	$d[S_2]/dt$
0	3.00	5.00	1.20	-1.20
0.2	3.24	4.76	0.98	-0.98
0.4	3.44	4.56	0.77	-0.77
0.6	3.59	4.41	0.59	-0.59
0.8	3.71	4.29	0.43	-0.43
1.0	3.80	4.20		

3.1.3.   Redo your simulation using the Fourth-Order Runge-Kutta.
Solution: The simulation result is shown below. Note that new values of $[S_1]$ and $[S_2]$ are determined using the equation:

$$([S_1], [S_2]) = ([S_1], [S_2]) + (0.2/6)(\alpha_1 + 2\alpha_2 + 2\alpha_3 + \alpha_4)$$

Time	S_1	S_2	α_1	α_2	α_3	α_4
0	3.00	5.00	(1.20,-1.20)	(1.10,-1.10)	(1.11,-1.11)	(1.00,-1.00)
0.2	3.22	4.78	(1.00,-1.00)	(0.90,-0.90)	(0.91,-0.91)	(0.81,-0.81)
0.4	3.40	4.60	(0.81,-0.81)	(0.72,-0.72)	(0.73,-0.73)	(0.64,-0.64)
0.6	3.55	4.45	(0.64,-0.64)	(0.56,-0.56)	(0.57,-0.57)	(0.50,-0.50)
0.8	3.66	5.34	(0.50,-0.50)	(0.43,-0.43)	(0.44,-0.44)	(0.38,-0.38)
1.0	3.75	4.25				

3.4   Derive an ODE model for the reaction-based model from Problem 1.3.
Solution: See Figure 8.4.

3.8   Using iBioSim or another tool that supports differential equation simulation, simulate the reaction-based model from Problem 1.3 with a Runge-Kutta method (ex. rkf45) with a time limit of 2100 and print interval of 50.

$$\frac{d[LacI]}{dt} = np\, k_o[S_1] - nc(K_r[P2][LacI]^{nc} - [S_5]) - k_d[LacI]$$

$$\frac{d[TetR]}{dt} = np\, k_o[S_2] - nc(K_r[P3][TetR]^{nc} - [S_6]) - k_d[TetR]$$

$$\frac{d[CI]}{dt} = np\, k_o[S_3] - 2(K_d[CI]^2 - [CI_2]) - k_d[CI]$$

$$\frac{d[CI_2]}{dt} = K_d[CI]^2 - [CI_2] - nc(K_r[P1][CI_2]^{nc} - [S_4])$$

$$\frac{d[P1]}{dt} = [S_1] - K_o[P1][RNAP] + [S_4] - K_r[P1][CI_2]^{nc}$$

$$\frac{d[P2]}{dt} = [S_2] - K_o[P2][RNAP] + [S_5] - K_r[P2][LacI]^{nc}$$

$$\frac{d[P3]}{dt} = [S_3] - K_o[P3][RNAP] + [S_6] - K_r[P3][TetR]^{nc}$$

$$\frac{d[RNAP]}{dt} = [S_1] - K_o[P1][RNAP] + [S_2] - K_o[P2][RNAP] +$$
$$[S_3] - K_o[P3][RNAP]$$

$$\frac{d[S_1]}{dt} = K_o[P1][RNAP] - [S_1]$$

$$\frac{d[S_2]}{dt} = K_o[P2][RNAP] - [S_2]$$

$$\frac{d[S_3]}{dt} = K_o[P3][RNAP] - [S_3]$$

$$\frac{d[S_4]}{dt} = K_r[P1][CI_2]^{nc} - [S_4]$$

$$\frac{d[S_5]}{dt} = K_r[P2][LacI]^{nc} - [S_5]$$

$$\frac{d[S_6]}{dt} = K_r[P3][TetR]^{nc} - [S_6]$$

FIGURE 8.4: ODE model for Problem 3.4.

Make a note of the simulation time and plot CI_2, LacI, and TetR. Next, simulate using Euler's method. Repeat the Euler simulation using different time steps until the results match up well. What time step is required for a good match? How do the simulation times compare?

Solution: The Runge-Kutta simulation of the reaction-based model from Problem 1.3 is shown in Figure 8.5. To get a comparable result using Euler's method requires using a maximum time step of about 0.01. The Runge-Kutta simulation time is about 50 percent larger than the simulation time for the Euler method (for example, 3 seconds for Runge-Kutta and 2 seconds for Euler).

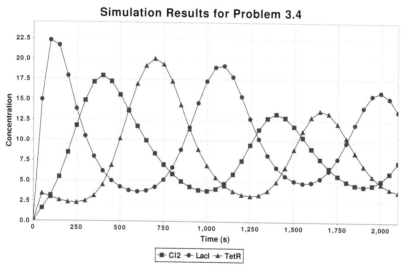

FIGURE 8.5: Simulation results for Problem 3.8.

4.1 Simulate by hand using Gillespie's SSA the following set of reactions assuming $(S_1, S_2) = (4, 4)$. Be sure to show all steps.

$$2S_1 \xrightarrow{0.1} 2S_2$$
$$S_1 + S_2 \xrightarrow{0.2} 2S_1$$

Create three separate trajectories with three reactions in each trajectory (if three are possible).

Solution: The solutions vary based on the random numbers selected. One possible set of three trajectories is shown below. The fourth trajectory is an example of one in which only two reactions can occur. This happens if the first reaction is selected twice in a row consuming all of the S_1 molecules.

Trajectory 1

Time	S_1	S_2	a_1	a_2	a_0	r_1	r_2
0	4	4	0.6	3.2	3.8	0.28	0.88
0.33	5	3	1.0	3.0	4	0.26	0.21
0.67	3	5	0.3	3.0	3.3	0.11	0.06
1.32	1	7					

Trajectory 2

Time	S_1	S_2	a_1	a_2	a_0	r_1	r_2
0	4	4	0.6	3.2	3.8	0.55	0.67
0.16	5	3	1.0	3.0	4	0.55	0.38
0.30	6	2	1.5	2.4	3.9	0.56	0.25
0.45	4	4					

Trajectory 3

Time	S_1	S_2	a_1	a_2	a_0	r_1	r_2
0	4	4	0.6	3.2	3.8	0.35	0.05
0.27	2	6	0.1	2.4	2.5	0.91	0.32
0.31	3	5	0.3	3.0	3.3	0.80	0.99
0.38	4	4					

Terminating Trajectory

Time	S_1	S_2	a_1	a_2	a_0	r_1	r_2
0	4	4	0.6	3.2	3.8	0.61	0.10
0.13	2	6	0.1	2.4	2.5	0.01	0.03
1.99	0	8					

4.3   In this problem, use a tool such as iBioSim that supports both ODE and SSA simulation. Simulate the reaction-based model from Problem 1.3 using ODE simulation with a time limit of 2100 and print interval of 50, and note the simulation time. Repeat using SSA simulation and a run count of 1. Again, note the simulation time. Plot the results for CI_2 for both the ODE and SSA simulations. Repeat the SSA simulation using a run count of 10, and plot the average for CI_2 with the ODE results for CI_2. Repeat again using a run count of 100. How do the simulation times compare as run counts increase? Do your ODE and SSA simulations agree as you increase the number of runs? Does ODE simulation appear to be reasonably accurate for this example?

Solution: The simulation results are shown in Figure 8.6. The runtime for one ODE run is about the same as 10 SSA runs (for example, 3 seconds for ODE, 0.5 seconds for one run of SSA, 3.7 seconds for 10 runs, and 36 seconds for 100). As the run counts increase, the average behavior smooths out, but the SSA simulation results are still quite different than the ODE simulation results. Therefore, the ODE simulation for this example is likely not very accurate.

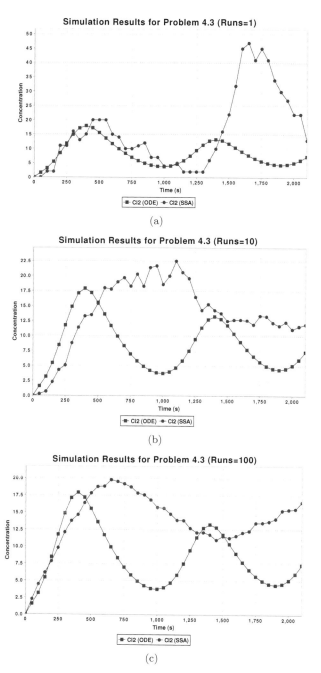

FIGURE 8.6:   Simulation results for Problem 4.3. Comparison of ODE simulation to SSA simulation with a run count of (a) 1, (b) 10, and (c) 100.

4.12 Create an SPN model for the reaction-based model from Problem 1.3.
Solution: See Figure 8.7.

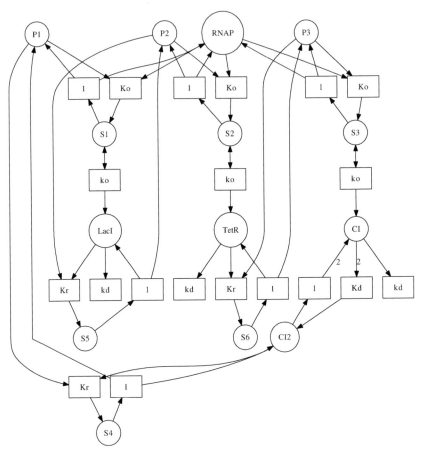

FIGURE 8.7: SPN for the reaction-based model from Problem 1.3.

5.1   Below are the chemical reactions involved in a competitive enzymatic reaction in which two substrates compete for a single enzyme.

$$E + S_1 \underset{k_2}{\overset{k_1}{\rightleftharpoons}} ES_1 \overset{k_3}{\rightarrow} E + P_1$$

$$E + S_2 \underset{k_5}{\overset{k_4}{\rightleftharpoons}} ES_2 \overset{k_6}{\rightarrow} E + P_2$$

5.1.1.   Using the law of mass action, write down the equations for the rates of change of $[S_1]$, $[S_2]$, $[ES_1]$, $[ES_2]$, $[P_1]$, and $[P_2]$.
Solution:

$$\frac{d[S_1]}{dt} = k_2[ES_1] - k_1[E][S_1] \tag{1}$$

$$\frac{d[S_2]}{dt} = k_5[ES_2] - k_4[E][S_2] \tag{2}$$

$$\frac{d[ES_1]}{dt} = k_1[E][S_1] - k_2[ES_1] - k_3[ES_1] \tag{3}$$

$$\frac{d[ES_2]}{dt} = k_4[E][S_2] - k_5[ES_2] - k_6[ES_2] \tag{4}$$

$$\frac{d[P_1]}{dt} = k_3[ES_1] \tag{5}$$

$$\frac{d[P_2]}{dt} = k_6[ES_2] \tag{6}$$

5.1.2.   If the amount of enzyme, E, is much less than either of the substrates, S_1 and S_2, then a steady state approximation can be used to simplify the equations for the rates of P_1 and P_2 formation. Using this approximation, derive an equation for $\frac{d[P_1]}{dt}$ in terms of the concentrations of the substrates, $[S_1]$ and $[S_2]$, the total enzyme concentration $[E_t]$, and the rate constants. You may also assume that $k_3 << k_2$ and that $k_6 << k_5$ to further simplify the derivations (hint: use this assumption at the beginning to simplify the rate equations for $[ES_1]$ and $[ES_2]$.
Solution: Since $k_3 << k_2$ and $k_6 << k_5$, Equations 3 and 4 can be reduced:

$$\frac{d[ES_1]}{dt} = k_1[E][S_1] - k_2[ES_1] \tag{7}$$

$$\frac{d[ES_2]}{dt} = k_4[E][S_2] - k_5[ES_2] \tag{8}$$

The total amount of enzyme can be defined to be:

$$[E_t] = [E] + [ES_1] + [ES_2] \tag{9}$$

$$[E] = [E_t] - [ES_1] - [ES_2] \tag{10}$$

Insert Equation 10 into Equations 7 and 8, and assuming that $\frac{d[ES_1]}{dt} \approx 0$ and $\frac{d[ES_2]}{dt} \approx 0$:

$$0 = k_1([E_t] - [ES_1] - [ES_2])[S_1] - k_2[ES_1] \tag{11}$$

$$0 = k_4([E_t] - [ES_1] - [ES_2])[S_2] - k_5[ES_2] \tag{12}$$

Rearrange Equation 12 to solve for $[ES_2]$:

$$[ES_2] = \frac{[S_2]([E_t] - [ES_1])}{[S_2] + \frac{k_5}{k_4}} \tag{13}$$

Inserting Equation 13 into Equation 11:

$$0 = k_1([E_t] - [ES_1] - \frac{[S_2]([E_t] - [ES_1])}{[S_2] + \frac{k_5}{k_4}})[S_1] - k_2[ES_1] \tag{14}$$

Multiply Equation 14 by $([S_2] + \frac{k_5}{k_4})$:

$$0 = k_1[S_1]\left([E_t]\left([S_2] + \frac{k_5}{k_4}\right) - [ES_1]\left([S_2] + \frac{k_5}{k_4}\right) - [S_2]([E_t] - [ES_1])\right)$$
$$-k_2[ES_1]\left([S_2] + \frac{k_5}{k_4}\right)$$

Result after cancelling some terms:

$$0 = k_1[S_1]\left([E_t]\left(\frac{k_5}{k_4}\right) - [ES_1]\left(\frac{k_5}{k_4}\right)\right) - k_2[ES_1]\left([S_2] + \frac{k_5}{k_4}\right) \tag{15}$$

Move $[ES_1]$ terms to left side:

$$[ES_1]\left(k_2\left([S_2] + \frac{k_5}{k_4}\right) + k_1[S_1]\left(\frac{k_5}{k_4}\right)\right) = k_1[S_1][E_t]\left(\frac{k_5}{k_4}\right) \tag{16}$$

Multiple Equation 16 by $\frac{k_4}{k_5 \cdot k_2}$:

$$[ES_1]\left(1 + \frac{k_4}{k_5}[S_2] + \frac{k_1}{k_2}[S_1]\right) = \frac{k_1}{k_2}[S_1][E_t] \tag{17}$$

Solve for $[ES_1]$:

$$[ES_1] = \frac{\frac{k_1}{k_2}[S_1][E_t]}{\left(1 + \frac{k_4}{k_5}[S_2] + \frac{k_1}{k_2}[S_1]\right)} \tag{18}$$

Insert Equation 18 into Equation 5:

$$\frac{d[P_1]}{dt} = \frac{k_3\frac{k_1}{k_2}[S_1][E_t]}{\left(1 + \frac{k_4}{k_5}[S_2] + \frac{k_1}{k_2}[S_1]\right)} \tag{19}$$

5.3 The P_{RE} promoter has the following four configurations:

s	O_1	O_2	Free energy
1	—	—	Reference
2	CII	—	ΔG_1
3	—	RNAP	ΔG_2
4	CII	RNAP	$\Delta G_1 + \Delta G_2 + \Delta G_{12}$

where ΔG_{12} represents the cooperative effect of CII bound to O_1 drawing RNAP to O_2. Assume the following experimental data:

DNA Template	O_1	O_2
wild type	12	30
O_1-	—	60
O_2-	85	—

where each entry is the number of nM before there is half occupation of the corresponding operator sites. For wild type, the data is the number of nM until half of the P_{RE} promoters have configuration 4. Using this experimental data, determine values for ΔG_1, ΔG_2, and ΔG_{12}. For RT, assume a value of 0.5961. (hint: Use O_1- and O_2- data first to calculate ΔG_2 and ΔG_1, respectively. Note that for these mutants not all configurations are possible. Also, note that the concentrations above are in nanomoles. Next, use the wild type data to find ΔG_{12}.)

Solution: Determine ΔG_2 using the O_1- data as follows:

$$f_3 = \frac{exp(-\Delta G_2/RT)[RNAP]}{1 + exp(-\Delta G_2/RT)[RNAP]}$$

$$0.5 = \frac{exp(-\Delta G_2/RT)60nM}{1 + exp(-\Delta G_2/RT)60nM}$$

$$1 = exp(-\Delta G_2/RT)60nM$$

$$\Delta G_2 = -RT\ln(\frac{1}{60nM})$$

$$\Delta G_2 = -9.91 \text{ kcal/M}$$

Determine ΔG_1 using the O_2- data as follows:

$$f_2 = \frac{exp(-\Delta G_1/RT)[CII]}{1 + exp(-\Delta G_1/RT)[CII]}$$

$$0.5 = \frac{exp(-\Delta G_1/RT)85nM}{1 + exp(-\Delta G_1/RT)85nM}$$

$$1 = exp(-\Delta G_1/RT)85nM$$

$$\Delta G_1 = -RT\ln(\frac{1}{85nM})$$

$$\Delta G_1 = -9.70 \text{ kcal/M}$$

Determine ΔG_{12} using wild type data as follows:

$$f_4 = \frac{exp(-(\Delta G_1 + \Delta G_2 + \Delta G_{12})/RT)[CII][RNAP]}{1 + c_1 + c_2 + exp(-(\Delta G_1 + \Delta G_2 + \Delta G_{12})/RT)[CII][RNAP]}$$

$$c_1 = exp(-\Delta G_1/RT)[CII]$$
$$= exp(9.70/RT)12nM$$
$$= 0.14$$

$$c_2 = exp(-\Delta G_2/RT)[RNAP]$$
$$= exp(9.91/RT)30nM$$
$$= 0.50$$

$$0.5 = \frac{exp(-\Delta G_{12} + 19.61)/RT)12nM \cdot 30nM}{1.64 + exp(-\Delta G_{12} + 19.61)/RT)12nM \cdot 30nM}$$

$$1.64 = exp(-\Delta G_{12} + 19.61)/RT)12nM \cdot 30nM$$

$$\Delta G_{12} = -RT\ln(\frac{1.64}{12nM \cdot 30nM}) + 19.61$$

$$\Delta G_{12} = -1.88 \text{ kcal/M}$$

5.7 Use abstraction to reduce the reaction-based model from Problem 1.3 and determine the simplified ODE model. To simplify the result, use the simpler dimerization abstraction in which the dimer species is rewritten in terms of the monomer species using Equation 5.42.

Solution: See Figure 8.8.

6.1 Consider the abstracted reaction-based model from Problem 5.7.

6.1.1. Determine the critical levels for the important species in this model assuming that the amplifier, a, is 0.95.

Solution:

$$\theta^1_{LacI} = \sqrt{a/(K_r - aK_r)} = 8.7$$
$$\theta^1_{TetR} = \sqrt{a/(K_r - aK_r)} = 8.7$$
$$\theta^1_{CI} = \sqrt[4]{a/(K_r K_d^2 - aK_r K_d^2)} = 13.2$$

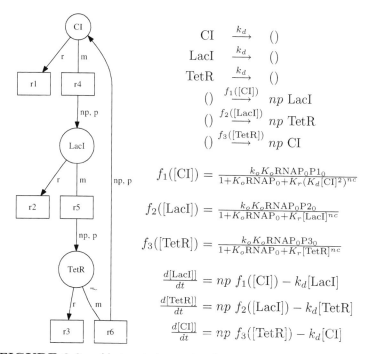

$$CI \xrightarrow{k_d} ()$$
$$LacI \xrightarrow{k_d} ()$$
$$TetR \xrightarrow{k_d} ()$$
$$() \xrightarrow{f_1([CI])} np \ LacI$$
$$() \xrightarrow{f_2([LacI])} np \ TetR$$
$$() \xrightarrow{f_3([TetR])} np \ CI$$

$$f_1([CI]) = \frac{k_o K_o \mathrm{RNAP_0 P1_0}}{1 + K_o \mathrm{RNAP_0} + K_r (K_d [CI]^2)^{nc}}$$

$$f_2([LacI]) = \frac{k_o K_o \mathrm{RNAP_0 P2_0}}{1 + K_o \mathrm{RNAP_0} + K_r [LacI]^{nc}}$$

$$f_3([TetR]) = \frac{k_o K_o \mathrm{RNAP_0 P3_0}}{1 + K_o \mathrm{RNAP_0} + K_r [TetR]^{nc}}$$

$$\frac{d[LacI]}{dt} = np \ f_1([CI]) - k_d[LacI]$$

$$\frac{d[TetR]}{dt} = np \ f_2([LacI]) - k_d[TetR]$$

$$\frac{d[CI]}{dt} = np \ f_3([TetR]) - k_d[CI]$$

FIGURE 8.8: Abstracted reaction-based model for Problem 5.7.

6.1.2.   Create a PLDE model and find the solution for each logical domain.
Solution:

$$\frac{d[LacI]}{dt} = k_o P1_0 (b_{CI} = 0) - k_d [LacI]$$

$$\frac{d[TetR]}{dt} = k_o P2_0 (b_{LacI} = 0) - k_d [TetR]$$

$$\frac{d[CI]}{dt} = k_o P3_0 (b_{TetR} = 0) - k_d [CI]$$

$$\Phi^{000} = \left(\frac{k_o P1_0}{k_d}, \frac{k_o P2_0}{k_d}, \frac{k_o P3_0}{k_d} \right) = (67, 67, 67)$$

$$\Phi^{001} = \left(0, \frac{k_o P2_0}{k_d}, \frac{k_o P3_0}{k_d} \right) = (0, 67, 67)$$

$$\Phi^{010} = \left(\frac{k_o P1_0}{k_d}, \frac{k_o P2_0}{k_d}, 0 \right) = (67, 67, 0)$$

$$\Phi^{011} = \left(0, \frac{k_o P2_0}{k_d}, 0 \right) = (0, 67, 0)$$

$$\Phi^{100} = \left(\frac{k_o P1_0}{k_d}, 0, \frac{k_o P3_0}{k_d} \right) = (67, 0, 67)$$

$$\Phi^{101} = \left(0, 0, \frac{k_o P3_0}{k_d} \right) = (0, 0, 67)$$

$$\Phi^{110} = \left(\frac{k_o P1_0}{k_d}, 0, 0 \right) = (67, 0, 0)$$

$$\Phi^{111} = (0, 0, 0)$$

6.1.3.   Create a stochastic FSM model.
Solution:

$$b_{CI} = 0 \wedge b_{LacI} = 0 \xrightarrow{q_1} b_{LacI} := 1$$

$$b_{CI} = 1 \wedge b_{LacI} = 0 \xrightarrow{q_2} b_{LacI} := 1$$

$$b_{LacI} = 1 \xrightarrow{q_3} b_{LacI} := 0$$

$$b_{LacI} = 0 \wedge b_{TetR} = 0 \xrightarrow{q_4} b_{TetR} := 1$$

$$b_{LacI} = 1 \wedge b_{TetR} = 0 \xrightarrow{q_5} b_{TetR} := 1$$

$$b_{TetR} = 1 \xrightarrow{q_6} b_{TetR} := 0$$

$$b_{TetR} = 0 \wedge b_{CI} = 0 \xrightarrow{q_7} b_{CI} := 1$$

$$b_{TetR} = 1 \wedge b_{CI} = 0 \xrightarrow{q_8} b_{CI} := 1$$

$$b_{CI} = 1 \xrightarrow{q_9} b_{CI} := 0$$

$$q_1 = np\ f_1(\theta^0_{CI})/\theta^1_{LacI} = 0.03$$
$$q_2 = np\ f_1(\theta^1_{CI})/\theta^1_{LacI} = 0.0014$$
$$q_3 = k_d\theta^1_{LacI}/\theta^1_{LacI} = 0.0075$$
$$q_4 = np\ f_2(\theta^0_{LacI})/\theta^1_{LacI} = 0.03$$
$$q_5 = np\ f_2(\theta^1_{LacI})/\theta^1_{LacI} = 0.0014$$
$$q_6 = k_d\theta^1_{TetR}/\theta^1_{TetR} = 0.0075$$
$$q_7 = np\ f_3(\theta^0_{TetR})/\theta^1_{CI} = 0.02$$
$$q_8 = np\ f_3(\theta^1_{TetR})/\theta^1_{CI} = 0.001$$
$$q_9 = k_d\theta^1_{CI}/\theta^1_{CI} = 0.0075$$

6.1.4. Analyze the stochastic FSM using Markov chain analysis. In particular, determine its CTMC and EMC representations. Then, using a computer program, find the steady-state distributions for the EMC and CTMC. Do your results make sense? Explain your answer.

Solution: The CTMC for the stochastic FSM is shown in Figure 8.9. Its transition rate matrix is given below:

$$Q = \begin{pmatrix} -0.08 & 0.02 & 0.03 & 0.0 & 0.03 & 0.0 & 0.0 & 0.0 \\ 0.0075 & -0.0389 & 0.0 & 0.03 & 0.0 & 0.0014 & 0.0 & 0.0 \\ 0.0075 & 0.0 & -0.0385 & 0.001 & 0.0 & 0.0 & 0.03 & 0.0 \\ 0.0 & 0.0075 & 0.0075 & -0.0164 & 0.0 & 0.0 & 0.0 & 0.0014 \\ 0.0075 & 0.0 & 0.0 & 0.0 & -0.0289 & 0.02 & 0.0014 & 0.0 \\ 0.0 & 0.0075 & 0.0 & 0.0 & 0.0075 & -0.0164 & 0.0 & 0.0014 \\ 0.0 & 0.0 & 0.0075 & 0.0 & 0.0075 & 0.0 & -0.016 & 0.001 \\ 0.0 & 0.0 & 0.0 & 0.0075 & 0.0 & 0.0075 & 0.0075 & -0.0225 \end{pmatrix}$$

The EMC for this CTMC is given below:

$$S = \begin{pmatrix} 0.0 & 0.25 & 0.375 & 0.0 & 0.375 & 0.0 & 0.0 & 0.0 \\ 0.1928 & 0.0 & 0.0 & 0.77121 & 0.0 & 0.03599 & 0.0 & 0.0 \\ 0.19481 & 0.0 & 0.0 & 0.02597 & 0.0 & 0.0 & 0.77922 & 0.0 \\ 0.0 & 0.45732 & 0.45732 & 0.0 & 0.0 & 0.0 & 0.0 & 0.08537 \\ 0.25952 & 0.0 & 0.0 & 0.0 & 0.0 & 0.69204 & 0.04844 & 0.0 \\ 0.0 & 0.45732 & 0.0 & 0.0 & 0.45732 & 0.0 & 0.0 & 0.08537 \\ 0.0 & 0.0 & 0.46875 & 0.0 & 0.46875 & 0.0 & 0.0 & 0.0625 \\ 0.0 & 0.0 & 0.0 & 0.33333 & 0.0 & 0.33333 & 0.33333 & 0.0 \end{pmatrix}$$

Note that this EMC has a period of 2, and it must be solved accordingly. The steady-state distribution for the EMC is:

$$\phi = (0.10, 0.14, 0.16, 0.12, 0.17, 0.13, 0.14, 0.03)$$

Using this result, the steady-state distribution for the CTMC is found to be:

$$\pi = (0.03, 0.09, 0.10, 0.18, 0.14, 0.19, 0.22, 0.03)$$

These results indicate that states 000 and 111 are the least likely which makes sense, since they are not valid states in an oscillator and should not persist. Also, states with two species above their critical threshold are more likely than those with only one. This also makes sense since production has higher rates than degradation.

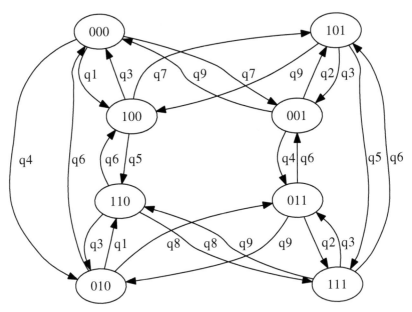

FIGURE 8.9: CTMC for Problem 6.1.4. State vector is $\langle \text{LacI}, \text{TetR}, \text{CI} \rangle$.

6.1.5. Construct a qualitative logical model and find its state graph. Do your results make sense? Explain your answer.

Solution: The qualitative logical model is shown below, and its state graph is shown in Figure 8.10.

$$b_{\text{LacI}} := (b_{CI} = 0)$$
$$b_{\text{TetR}} := (b_{LacI} = 0)$$
$$b_{\text{CI}} := (b_{TetR} = 0)$$

This model indicates that states 000 and 111 are transient which makes sense, since they are not valid states in an oscillator. Once a valid oscillator state is reached, the genetic circuit moves through an oscillation cycle.

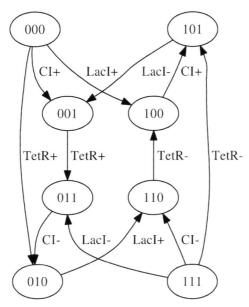

FIGURE 8.10: State graph for Problem 6.1.5. State vector is
⟨LacI, TetR, CI⟩.

7.1   An XOR gate is one in which its output goes high when exactly one of its inputs is high. The behavior of an XOR gate is defined by the following truth table:

A	B	C
0	0	0
0	1	1
1	0	1
1	1	0

7.1.1.   Design and simulate a genetic XOR gate.
Solution: See Figure 8.11.
7.1.2.   Design a PoPS XOR gate.
Solution: See Figure 8.12.

7.3   Design and simulate a sequential circuit with one input and one output that toggles its output every time it receives a pulse on its input. In other words, the output goes high after the input goes high, and it stays high until the input goes back to low and then to high again.
Solution: See Figure 8.13.

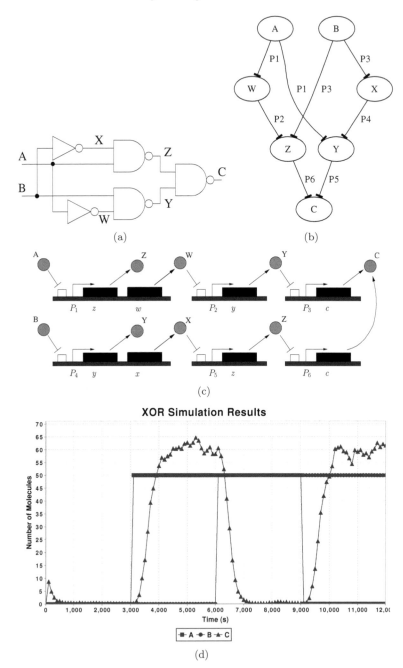

(a)

(b)

(c)

(d)

FIGURE 8.11: A genetic XOR gate. (a) Logic diagram.
(b) Genetic circuit model. (c) Genetic implementations.
(d) Average of 100 stochastic simulation runs.

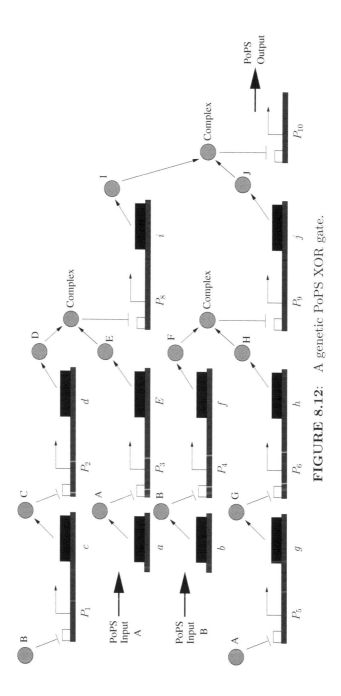

FIGURE 8.12: A genetic PoPS XOR gate.

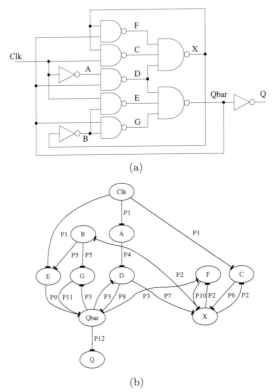

(a)

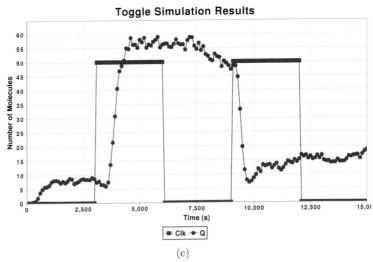

(b)

Toggle Simulation Results

(c)

FIGURE 8.13: A genetic toggle. (a) Logic diagram. (b) Genetic circuit model. (c) Average of 100 stochastic simulation runs.

References

Abrash, H. (1986). Studies concerning affinity. *Journal of Chemical Education*, **63**, 1044–1047. English translation of Waage and Guldberg's 1864 paper.

Ackers, G. K., Johnson, A. D., and Shea, M. A. (1982). Quantitative model for gene regulation by lambda phage repressor. *Proceedings of the National Academy of Sciences USA*, **79**(4), 1129–1133.

Alberghina, L. and Westerhoff, H. V. (2005). *Systems Biology: Definitions and Perspectives*. Birkhauser.

Alberts, B., Johnson, A., Lewis, J., Raff, M., Roberts, K., and Walter, P. (2002). *Molecular Biology of the Cell*. Garland Science.

Alon, U. (2007). *An Introduction to Systems Biology: Design Principles of Biological Circuits*. CRC Press.

Anderson, J. C., Clarke, E. J., and Arkin, A. P. (2006). Environmentally controlled invasion of cancer cells by engineering bacteria. *Journal of Molecular Biology*, **355**, 619–627.

Arkin, A. (2000). Signal processing by biochemical reaction networks. In J. Walleczek, editor, *Self-Organized Biological Dynamics and Nonlinear Control*, chapter 5. Cambridge University Press.

Arkin, A. (2008). Setting the standard in synthetic biology. *Nature Biotechnology*, **26**, 771–774.

Arkin, A., Ross, J., and McAdams, H. (1998). Stochastic kinetic analysis of developmental pathway bifurcation in phage lambda-infected escherichia coli cells. *Genetics*, **149**, 1633–1648.

Baldi, P. and Hatfield, G. W. (2002). *DNA Microarrays and Gene Expression*. Cambridge University Press.

Barker, N. (2007). *Learning Genetic Regulatory Network Connectivity from Time Series Data*. Ph.D. thesis, University of Utah.

Barker, N., Myers, C., and Kuwahara, H. (2006). Learning genetic regulatory network connectivity from time series data. *Advances in Applied Artifical Intelligence*, **4031**, 962–971.

Beal, M. J., Falciani, F., Ghahramani, Z., Rangel, C., and Wild, D. L. (2005). Bayesian approach to reconstructing genetic regulatory networks with hidden factors. *Bioinformatics*, **21**(3), 349–56.

Berg, J., Tymoczko, J., and Stryer, L. (2002). *Biochemistry*. W. H. Freeman and Co.

Bernard, A. and Hartemink, A. J. (2005). Informative structure priors: joint learning of dynamic regulatory networks from multiple types of data. In *Pacific Symposium on Biocomputing*, pages 459–70.

Borisuk, M. T. and Tyson, J. J. (1998). Bifurcation analysis of a model of mitotic control in frog eggs. *Journal of Theoretical Biology*, **195**, 69–85.

Brazil, G. M., Kenefick, L., Callanan, M., Haro, A., de Lorenzo, V., Dowling, D. N., and O'Gara, F. (1995). Construction of a rhizosphere pseudomonad with potential to degrade polychlorinated biphenyls and detection of *bph* gene expression in the rhizosphere. *Applied and Environmental Microbiology*, **61**(5), 1946–1952.

Briggs, G. E. and Haldane, J. B. S. (1925). A note on the kinetics of enzyme action. *Biochemical Journal*, **19**, 339–339.

Brown, P. A. and Botstein, D. (1999). Exploring the new world of the genome with DNA microarrays. *Nature Genetics*, **21**(suppl.), 33–37.

Bunow, B., Kernevez, J.-P., Joly, G., and Thomas, D. (1980). Pattern formation by reaction-diffusion instabilities: Application to morphogenesis in drosophila. *Journal of Theoretical Biology*, **84**, 629–649.

Canton, B., Labno, A., and Endy, D. (2008). Refinement and standardization of synthetic biological parts and devices. *Nature Biotechnology*, **26**, 787–793.

Cao, Y., Gillespie, D. T., and Petzold, L. R. (2006). Efficient step size selection for the tau-leaping simulation method. *Journal of Chemical Physics*, **124**(4), 044109.

Cases, I. and de Lorenzo, V. (2005). Genetically modified organisms for the environment: stories of success and failure and what we have learned from them. *International Microbiology*, **8**, 213–222.

Chen, K. C., Csikász-Nagy, A., Györffy, B., Val, J., Novak, B., and Tyson, J. J. (2000). Kinetic analysis of a molecular model of the budding yeast cell cycle. *Molecular Biology of the Cell*, **11**(1), 369–391.

Chien, C. T., Bartel, P. L., Sternglanz, R., and Fields, S. (1991). The two-hybrid system: a method to identify and clone genes for proteins that interact with a protein of interest. *Proceedings of the National Academy of Sciences USA*, **88**(21), 9578–82.

D'haeseleer, P., Liang, S., and Somogyi, R. (2000). Genetic network inference: from co-expression clustering to reverse engineering. *Bioinformatics*, **16**(8), 707–726.

Eisen, M. B., Spellman, P. T., Browndagger, P. O., and Botstein, D. (1998). Cluster analysis and display of genome-wide expression patterns. *Proceedings of the National Academy of Sciences USA*, **95**, 14863–14868.

Elowitz, M. and Leibler, S. (2000). A synthetic oscillatory network of transcriptional regulators. *Nature*, **403**(6767), 335–338.

Endy, D. (2005). Foundations for engineering biology. *Nature*, **438**, 449–453.

Endy, D., You, L., Yin, J., and Molineux, I. J. (2000). Computation, prediction, and experimental tests of fitness for bacteriophage t7 mutants with permuted genomes. *Proceedings of the National Academy of Sciences USA*, **97**(10), 5375–5380.

Friedman, N., Linial, M., Nachman, I., and Pe'er, D. (2000). Using bayesian networks to analyze expression data. *Journal of Computational Biology*, **7**(3–4), 601–620.

Gardner, T. S., Cantor, C. R., and Collins, J. J. (2000). Construction of a genetic toggle switch in *escherichia coli*. *Nature*, **403**, 339–342.

Gibson, M. and Bruck, J. (2000). Efficient exact stochastic simulation of chemical systems with many species and many channels. *Journal of Physical Chemistry*, **A 104**, 1876–1889.

Gillespie, D. T. (1977). Exact stochastic simulation of coupled chemical reactions. *Journal of Physical Chemistry*, **81**(25), 2340–2361.

Gillespie, D. T. (1992). *Markov Processes: An Introduction for Physical Scientists*. Academic Press.

Gillespie, D. T. (2005). *Stochastic Chemical Kinetics*, chapter 5.11, pages 1735–1752. Springer-Verlag. In Handbook of Materials Modeling (edited by S. Yip).

Gillespie, D. T. and Petzold, L. R. (2003). Tau leaping. *Journal of Chemical Physics*, **119**, 8229–8234.

Giona, M. and Adrover, A. (2002). Modified model for the regulation of the tryptophan operon in escherichia coli. *Biotechnology & Bioengineering*, **80**(3), 297–304.

Glass, L. (1977). *Global analysis of nonlinear chemical kinetics*, pages 311–349. Plenum Press, New York. In Statistical Mechanics, Part B: Time Dependent Processes (edited by B. Berne).

Gonick, L. and Wheelis, M. (1983). *The Cartoon Guide to Genetics*. HarperPerennial.

Goodwin, B. C. (1963). *Temporal Organization of Cells*. Academic Press.

Goodwin, B. C. (1965). Oscillatory behavior in enzymatic control processes. *Advances in Enzyme Regulation*, pages 425–438.

Goodwin, B. C. and Kauffman, S. A. (1990). Spatial harmonics and pattern specification in early drosophila development: Part I. bifurcation sequences and gene expression. *Journal of Theoretical Biology*, **144**, 303–319.

Goss, P. J. and Peccoud, J. (1998). Quantitative modeling of stochastic systems in molecular biology by using stochastic Petri nets. *Proceedings of the National Academy of Sciences USA*, **95**(12), 6750–6755.

Gottesman, M. and Weisberg, R. (2004). Little lambda, who made thee? *Microbiology and Molecular Biology Reviews*, **68**(4), 796–813.

Gouzé, J.-L. and Sari, T. (2001). A class of piecewise linear differential equations arising in biological models. Technical report, INRIA Sophia-Antipolis, Sophia-Antipolis. Technical Report RR-4207.

Guet, C., Elowitz, M., Hsing, W., and Leibler, S. (2002). Combinatorial synthesis of genetic networks. *Science*, **296**, 1466–1470.

Hartemink, A. J., Gifford, D. K., Jaakkola, T. S., and Young, R. A. (2001). Using graphical models and genomic expression data to statistically validate models of genetic regulatory networks. In *Pacific Symposium on Biocomputing*, pages 422–33.

Hasty, J., McMillen, D., and Collins, J. J. (2002). Engineered gene circuits. *Nature*, **420**, 224–230.

Heckerman, D. (1996). A tutorial on learning with bayesian networks. Technical report, Microsoft Research, Microsoft Corporation, One Microsoft Way, Redmond, WA 98052.

Henri, V. (1903). Lois générales de l'action des diastases. Hermann, Paris.

Hoek, M. J. V. and Hogeweg, P. (2006). In silico evolved lac operons exhibit bistability for artificial inducers, but not for lactose. *Biophysical Journal*, **91**, 2833–2843.

Hucka, M., Finney, A., Sauro, H. M., Bolouri, H., Doyle, J. C., Kitano, H., Arkin, A. P., Bornstein, B. J., Bray, D., Cornish-Bowden, A., Cuellar, A. A., Dronov, S., Gilles, E. D., Ginkel, M., Gor, V., Goryanin, I. I., Hedley, W. J., Hodgman, T. C., Hofmeyr, J. H., Hunter, P. J., Juty, N. S., Kasberger, J. L., Kremling, A., Kummer, U., Le Novère, N., Loew, L. M., Lucio, D., Mendes, P., Minch, E., Mjolsness, E. D., Nakayama, Y., Nelson, M. R., Nielsen, P. F., Sakurada, T., Schaff, J. C., Shapiro, B. E., Shimizu, T. S., Spence, H. D., Stelling, J., Takahashi, K., Tomita, M., Wagner, J., and Wang, J. (2003). The systems biology markup language (SBML): a medium for representation and exchange of biochemical network models. *Bioinformatics*, **19**(4), 524–531.

Husmeier, D. (2003). Sensitivity and specificity of inferring genetic regulatory interactions from microarray experiments with dynamic bayesian networks. *Bioinformatics*, **19**(17), 2271–82.

International Human Genome Sequencing Consortium (2001). Initial sequencing and analysis of the human genome. *Nature*, **409**, 860–921.

Jacob, F. and Monod, J. (1961). Genetic regulatory mechanisms in the synthesis of proteins. *Journal of Molecular Biology*, **3**, 318–356.

Johnson, A. D., Meyer, B. J., and Ptashne, M. (1979). Interactions between DNA-bound repressors govern regulation by the λ phage repressor. *Proceedings of the National Academy of Sciences USA*, **76**, 5061–5065.

Johnson, S. C. (1967). Hierarchical clustering schemes. *Psychometrika*, **2**, 241–254.

Jong, H. D., Page, M., Hernandez, C., and Geiselmann, J. (2001). Qualitative simulation of genetic regulatory networks: Method and application. In *Proc. 17th Int. Joint Conf. Artif. Intell. (IJCAI-01)*, pages 67–73. Morgan Kaufmann.

Kahn, P. (1995). From genome to proteome: Looking at cell's proteins. *Science*, **270**, 369–370.

Kauffman, S. A. (1969). Metabolic stability and epigenesis in randomly constructed genetic nets. *Journal of Theoretical Biology*, **22**, 437–467.

Kauffman, S. A., Shymko, R. M., and Trabert, K. (1978). Control of sequential compartment formation in Drosophila. *Science*, **199**, 259–270.

King, R. and Stansfield, W. (1990). *A Dictionary of Genetics*. Oxford University Press.

Kitano, H. (2001). *Foundations of Systems Biology*. MIT Press.

Konopka, A. (2007). *Systems Biology: Principles, Methods, and Concepts*. CRC Press.

Kuwahara, H. (2007). *Model Abstraction and Temporal Behavior Analysis of Genetic Regulatory Networks*. Ph.D. thesis, University of Utah.

Kuwahara, H. and Mura, I. (2008). An efficient and exact stochastic simulation method to analyze rare events in biochemical systems. *The Journal of Chemical Physics*, **129**(16).

Kuwahara, H. and Myers, C. (2008). Production-passage time approximation: A new approximation method to accelerate the simulation process of enzymatic reactions. *Journal of Computational Biology*, **15**(7), 779–792.

Kuwahara, H., Myers, C., Barker, N., Samoilov, M., and Arkin, A. (2006). Automated abstraction methodology for genetic regulatory networks. *Transactions on Computational Systems Biology*, **VI**, 150–175.

Lacalli, T. C. (1990). Modeling the drosophila pair-rule pattern by reaction diffusion: Gap input and pattern control in a 4-morphogen system. *Journal of Theoretical Biology*, **144**, 171–194.

Lederberg, E. (1951). Lysogenicity in e. coli k-12. *Microbial Genetic Bulletin*, **36**(560).

Leloup, J. C. and Goldbeter, A. (1951). A model for the circadian rhythms in drosophila incorporating the formation of a complex between the per and tim proteins. *Journal of Biological Rhythms*, **13**(1), 70–87.

Lipschutz, R. J., Fodor, S. P. A., Gingeras, T. R., and Lockhart, D. J. (1999). High density synthetic oligonucleotide arrays. *Nature Genetics*, **21**(suppl.), 20–24.

Little, S., Seegmiller, N., Walter, D., Myers, C. J., and Yoneda, T. (2006). Verification of analog/mixed-signal circuits using labeled hybrid Petri nets. In *Proceedings of the International Conference on Computer-Aided Design*, pages 275–282.

Lockhart, D. J. and Winzeler, E. A. (2000). Genomics, gene expression, and DNA arrays. *Nature*, **405**, 827–836.

Lwoff, A. and Gutmann, A. (1950). Recherches sur un *Bacillus megatherium* lysogéne. *Annuals Institute Pasteur (Paris)*, **78**, 7111–739.

MacQueen, J. B. (1967). Some methods for classification and analysis of multivariate observations. *Proceedings of 5th Berkeley Symposium on Mathematical Statistics and Probability, Berkeley*, **1**, 281–297.

Mann, M. (1999). Quantitative proteomics. *Nature Biotechnology*, **17**, 954–955.

Marsan, M., Conte, G., and Balbo, G. (1984). A class of generalized stochastic Petri nets for the performance evaluation of multiprocessor systems. *ACM Transactions on Computer Systems*, **2**(2), 93–122.

Marsan, M., Balbo, G., Conte, G., Donatelli, S., and Franceschinis, G. (1994). *Modelling with Generalized Stochastic Petri Nets*. John Wiley & Sons, Inc., New York, NY, USA.

Matsuno, H., Doi, A., Nagasaki, M., and Miyano, S. (2000). Hybrid Petri net representation of gene regulatory network. In R. B. Altman, A. K. Dunker, L. Hunter, and T. E. Klein, editors, *Pacific Symposium on Biocomputing*, volume 5, pages 341–352, Singapore. World Scientific Press.

McAdams, H. H. and Shapiro, L. (1995). Circuit simulation of genetic networks. *Science*, **269**(5224), 650–656.

Michaelis, L. and Menten, M. (1913). Die kinetik der invertinwirkung. *Biochemische Zeitschrift*, **49**, 333–369.

Molloy, M. K. (1982). Performance analysis using stochastic Petri nets. *IEEE Transactions on Computers*, **31**(9), 913–917.

Myasnikova, E., Samsonova, A., Kozlov, K., Samsonova, M., and Reinitz, J. (2001). Registration of the expression patterns of drosophila segmentation genes by two independent methods. *Bioinformatics*, **17**(1), 3–12.

Myers, C. J. (2001). *Asynchronous Circuit Design*. John Wiley & Sons.

Nachman, I., Regev, A., and Friedman, N. (2004). Inferring quantitative models of regulatory networks from expression data. *Bioinformatics*, **20 Suppl 1**, 148–1256.

Nguyen, N. (2008). *Design and Analysis of Genetic Circuits*. Master's thesis, University of Utah.

Nguyen, N., Kuwahara, H., Myers, C., and Keener, J. (2007). The design of a genetic Muller C-element. In *The 13th IEEE International Symposium on Asynchronous Circuits and Systems*, pages 95–104.

Novak, B. and Tyson, J. J. (1995). Quantitative analysis of a molecular model of mitotic control in fission yeast. *Journal of Theoretical Biology*, **173**, 283–305.

Novak, B., Csikasz-Nagy, A., Gyorffy, B., Chen, K. C., and Tyson, J. J. (1998). Mathematical model of the fission yeast cell cycle with checkpoint controls at the g1/s, g2/m and metaphase/anaphase transitions. *Biophysical Chemistry*, **72**, 185–200.

Ong, I. M., Glasner, J. D., and Page, D. (2002). Modelling regulatory pathways in e. coli from time series experssion profiles. *Bioinformatics*, **18**, s241–s248.

Osterhout, R. E., Figueroa, I. A., Keasling, J. D., and Arkin, A. P. (2007). Global analysis of host response to induction of a latent bacteriophage. *BMC Microbiology*, **7**(82).

Palsson, B. (2006). *Systems Biology: Properties of Reconstructed Networks*. Cambridge University Press.

Pandey, A. and Mann, M. (2000). Proteomics to study genes and genomes. *Nature*, **405**, 837–846.

Pe'er, D. (2005). Bayesian network analysis of signaling networks: a primer. *Science's Signal Transduction Knowledge Environment*, **2005**(281), pl4.

Plahte, E., Mestl, T., and Omholt, S. W. (1994). Global analysis of steady points for systems of differential equations with sigmoid interactions. *Dynamics and Stability of Systems*, **9**(4), 275–291.

Press, W. H., Flannery, B. P., Teukolsky, S. A., and Vetterling, W. T. (1992). *Numerical Recipes in C: The art of Scientic Computing, 2nd ed.* Cambridge University Press.

Ptashne, M. (1992). *A Genetic Switch*. Cell Press & Blackwell Scientific Publishing.

Rao, C. V. and Arkin, A. P. (2003). Stochastic chemical kinetics and the quasi-steady-state assumption: Application to the gillespie algorithm. *Journal of Physical Chemistry*, **118**(11), 4999–5010.

Rathinam, M., Petzold, L., Cao, Y., and Gillespie, D. (2003). Implicit tau leaping. *Journal of Chemical Physics*, **119**, 12784–12794.

Reinitz, J. and Vaisnys, J. R. (1990). Theoretical and experimental analysis of the phage lambda genetic switch implies missing levels of cooperativity. *Journal of Theoretical Biology*, **145**, 295–318.

Ren, B. and Dynlacht, B. D. (2004). Use of chromotin immunoprecipitation assays in genome-wide location analysis of mammalian transcription factors. *Methods in Enzymology*, **376**, 304–15.

Ro, D.-K., Paradise, E. M., Ouellet, M., Fisher, K. J., Newman, K. L., Ndungu, J. M., Ho, K. A., Eachus, R. A., Ham, T. S., Kirby, J., Chang, M. C. Y., Withers, S. T., Shiba, Y., Sarpong, R., and Keasling, J. D. (2006). Production of the antimalarial drug precursor artemisinic acid in engineered yeast. *Nature*, **440**, 940–943.

Ruoff, P., Vinsjevik, M., Monnerjahn, C., and Rensing, L. (2001). The Goodwin model : Simulating the effect of light pulses on the circadian sporulation rhythm of neurospora crassa. *Journal of Theoretical Biology*, **209**, 29–42.

Sachs, K., Perez, O., Pe'er, D., Lauffenburger, D., and Nolan, G. (2005). Causal protein-signaling networks derived from multiparameter single-cell data. *Science*, **308**, 523–529.

Santillán, M. and Mackey, M. C. (2004). Influence of catabolite repression and inducer exclusion on the bistable behavior of the lac operon. *Biophysical Journal*, **86**, 1282–1292.

Santillán, M. and Zeron, E. S. (2004). Dynamic influence of feedback enzyme inhibition and transcription attenuation on the tryptophan operon response to nutritional shifts. *Journal of Theoretical Biology*, **231**(2), 287–298.

Santillán, M. and Zeron, E. S. (2006). Analytical study of the multiplicity of regulatory mechanisms in the tryptophan operon. *Bulletin of Mathematical Biology*, **68**(2), 343–359.

Sarpeshkar, R. (2006). Brain power. *IEEE Spectrum*, pages 24–29.

Segel, L. A. and Slemrod, M. (1989). The quasi-steady-state assumption: A case study in perturbation. *SIAM Review*, **31**(3), 446–477.

Shea, M. A. and Ackers, G. K. (1985). The OR control system of bacteriophage lambda: a physical-chemical model for gene regulation. *Journal of Molecular Biology*, **181**, 211–230.

Sprinzak, D. and Elowitz, M. B. (2005). Reconstruction of genetic circuits. *Nature*, **438**, 443–448.

Stewart, W. J. (1994). *Introduction to the Numerical Solution of Markov Chains*. Priceton University Press.

Strogatz, S. H. (1994). *Nonlinear Dynamics and Chaos: With Applications to Physics, Biology, Chemistry, and Engineering*. Westview Press.

Stundzia, A. B. and Lumsden, C. J. (1996). Stochastic simulation of coupled reaction-diffusion processes. *Journal of Computational Physics*, **127**, 196–207.

Takahashi, K., Arjunan, S. N. V., and Tomita, M. (2005). Space in systems biology of signaling pathways–towards intracellular molecular crowding in silico. *FEBS Letters*, **579**, 1783–1788.

Tavazoie, S., Hughes, J. D., Campbell, M. J., Cho, R. J., and Church, G. M. (1999). Systematic determination of genetic network architecture. *Nature Genetics*, **22**, 281–285.

Thieffry, D. and Thomas, R. (1995). Dynamical behaviour of biological networks: II. Immunity control in bacteriophage lamabda. *Bulletin of Mathematical Biology*, **57**(2), 277–297.

Thomas, R. (1991). Regulatory metworks seen as asynchronous automata: A logical description. *Journal of Theoretical Biology*, **153**, 1–23.

Tozeren, A. and Byers, S. (2003). *New Biology for Engineers and Computer Scientists*. Prentice Hall.

Tryon, R. C. (1939). *Cluster Analysis*. New York: McGraw-Hill.

Turing, A. (1951). The chemical basis of morphogenesis. *Philosophical Transactions of the Royal Society*, **B. 273**, 37–72.

Tyson, J. (1999). Models of cell cycle control in eukaryotes. *Journal of Biotechnology*, **71**(1–3), 239–244.

Tyson, J. and Othmer, H. (1978). The dynamics of feedback control circuits in biochemical pathways. *Progress in Theoretical Biology*, **5**, 1–62.

Tyson, J., Novak, B., Odell, G. M., Chen, K., and Thron, C. D. (1996). Chemical kinetic theory: Understanding cell-cycle regulation. *Trends Biochemical Sciences*, **21**(3), 89–96.

Ueda, H. R., Hagiwara, M., and Kitano, H. (2001). Robust oscillations within the interlocked feedback model of drosophila circadian rhythm. *Journal of Theoretical Biology*, **210**, 401–406.

Venter, J. C. *et al.* (2001). The sequence of the human genome. *Science*, **291**, 1304–1351.

Vilar, J. M., Guet, C. C., and Leibler, S. (2003). Modeling network dynamics: the lac operon, a case study. *Journal of Cell Biology*, **161**, 471–476.

Waage, P. and Guldberg, C. M. (1864). Studies concerning affinity. *Forhandlinger: Videnskabs - Selskabet i Christinia*, **35**.

Watson, J., Baker, T., Bell, S., Gann, A., Levine, M., and Losick, R. (2003). *Molecular Biology of the Gene*. Benjamin Cummings.

Watson, J. D. and Crick, F. H. C. (1953). A structure for deoxyribose nucleic acid. *Nature*, **171**, 737–738.

Wilkinson, D. J. (2006). *Stochastic Modelling for Systems Biology*. CRC Press.

Wolf, D. M. and Arkin, A. P. (2002). Fifteen minutes of *fim*: Control of type 1 pili expression in *e. coli*. *OMICS: A Journal of Integrative Biology*, **6**(1), 91–114.

Wong, P., Gladney, S., and Keasling, J. D. (1997). Mathematical model of the lac operon: inducer exclusion, atabolite repression, and diauxic growth on glucose and lactose. *Biotechnology Progress*, **13**, 132–143.

Wright, M. (2004). *Introduction to Chemical Kinetics*. John Wiley and Sons.

Xiu, Z. L., Chang, Z. Y., and Zeng, A. P. (2002). Nonlinear dynamics of regulation of bacterial trp operon: model analysis of integrated effects of repression, feedback inhibition, and attenuation. *Biotechnology Progress*, **18**(4), 686–93.

Yildirim, N. and Mackey, M. C. (2003). Feedback regulation in the lactose operon: a mathematical modeling study and comparison with experimental data. *Biophysical Journal*, **84**, 2841–2851.

Yildirim, N., Santillán, M., Horike, D., and Mackey, M. C. (2004). Dynamics and bistability in a reduced model of the lac operon. *Chaos*, **14**, 279–292.

You, L. and Yin, J. (2006). Evolutionary design on a budget: robustness and optimality of bacteriophage t7. *Systems Biology*, **153**(2), 46–52.

Yu, J., Smith, V. A., Wang, P. P., Hartemink, A. J., and Jarvis, E. D. (2004). Advances to bayesian network inference for generating causal networks from observational biological data. *Bioinformatics*, **20**, 3594–3603.

Zhu, H. and Snyder, M. (2001). Protein arrays and microarrays. *Current Opinion in Chemical Biology*, **5**, 40–45.

Glossary

3' end The end of the DNA that terminates in a 3' hydroxyl group (-OH). The genetic information is always read from the 5' to the 3' end, 5

5' end The end of the DNA that terminates in a 5' phosphate group (-PO4). The genetic information is always read from the 5' to the 3' end, 5

A site The location on the large subunit of the ribosome which binds to a new tRNA, 15

absolute error level The desired maximum acceptable error, 90

abstraction The engineering principle of using high-level models that hide low-level details to make the design process more efficient, 187

acceptor site Location on the tRNA that binds to the specific amino acid for the codon that is associated with the tRNA, 15

accuracy control parameter A parameter used in tau-leaping to indicate the amount of change of a propensity function that is allowed during one τ time step, 116

activate The binding of proteins to an operator site to turn on transcription of a gene, 13

activated rate The rate of transcription of a gene once it becomes activated by, for example, the binding of a transcription factor to an associated operator site. See also basal rate, 20

activating domain (AD) The portion of a eukaryotic transcription factor that is responsible for activating transcription, 55

activation energy An energy barrier that must be overcome before a reaction can occur, 3

activator A transcription factor that enhances transcription of a gene usually by helping RNAP bind to the promoter, 13

adaptive stepsize control A process that changes the stepsize dynamically during a numerical simulation to slow down when rates are changing rapidly and speed up when they are changing slowly, 90

adenine The chemical base "A" found in DNA and RNA, 5

affinity The attraction between two molecules. For example, how strongly a transcription factor is drawn to an operator site, 20

alternative splicing The process in which different portions of an mRNA sequence are considered introns. The result is one mRNA sequence from one gene can code for multiple proteins, 9

amino acids The molecules that make up proteins. There are 20 different kinds used by living organisms, 5

analog circuit A circuit having continuous valued states, 163

AND gate A combinational logic gate that has its output go high only when all of its inputs are high, 193

anhydrotetracycline (aTc) A small molecule chemical inducer which binds to the protein TetR preventing it from being able to act as a transcription factor, 195

animal viruses Viruses that infect animals and humans, 16

anti-codon site Location on a tRNA that binds to a particular codon, 15

anti-terminator A protein that when bound to RNAP prevents the RNAP from falling off the DNA when it encounters a terminator switch, 30

antisense strand See template strand, 13

aperiodic state A state which is not periodic (i.e., $p = 1$), 175

assembly The fourth stage of virus replication in which the new viruses are constructed within the cell either by chance or with the assistance of molecular chaperons, 16

atoms The basic building block for all matter whether living or not, 1

ATP *Adenosine triphosphate*, also known as the universal energy currency of living organisms, 3

attachment The first stage of virus replication in which the virus binds to the host's cell wall, 16

attachment site The location where the phage λ DNA is split when it is integrated within the host's chromosome, 29

attp site The location of the attachment site on the phage λ genome, 29

automated construction The engineering principle of separating design issues from construction issues, 187

automated sequencing An experimental process for finding the DNA sequence for a strand of DNA, 187

autoregulation The process in which a protein modifies its own rate of production, 198

average phage input The proportion of phages to bacteria in a population, 122

background knowledge Information about a genetic circuit that is known before learning from experimental data, 70

backward Euler method A version of Euler's method which cannot determine the rates of change directly but rather it must instead solve a set of algebraic equations, 88

bacteria A single-celled organism, 9

bacteriophages Viruses that only infect bacteria, 16

bait The protein being studied, 56

basal rate The basic rate of transcription without the gene being activated. See also activated rate, 20

base bin assignment A partial bin assignment which is the point of comparison for all other partial bin assignments, 72

Bayesian networks A directed acyclic graph representation of a joint probability distribution, 51

Bayesian learning Methods to infer the most likely Bayesian network that may have produced a given set of experimental data, 63

Bayesian scoring metric A method of assigning a score to a Bayesian network used to determine the best network to explain the data, 64

bifurcation diagram A diagram that shows how stability changes with respect to a parameter change, 93

bifurcation point A parameter value in which a substantial change in stability occurs in the phase space, 92

bimolecular reaction channel A reaction that has two reactants, 104

bin assignment An assignment to a particular bin for each species, 57

binding domain (BD) The portion of a eukaryotic transcription factor that is responsible for binding to the operator site, 55

bins A discretization of experimental data where data assigned to a particular bin has a value between two levels which define the bin, 57

BioBricks^{TM} A standard used for biological parts, 188

bioinformatics The use of information technology to solve problems in molecular biology. Typically, it involves the analysis of static data such as DNA sequences, xxiii

biological databases An electronic collection of biological data that can be easily updated, queried, and retrieved, xxv

bistability Having two stable states, 198

black boundary A boundary between two domains in a piecewise differential equation model in which all trajectories leave both domains from this boundary, 167

bootstrap A method where data is divided in several different ways and a network is learned for each grouping of data. Statistical confidence in a feature can be determined by determining how likely it is to find the feature in one of the learned networks, 67

capsule Outer surface of a prokaryotic cell, 10

carbohydrates Macromolecules made up of carbon and water ($C_n(H_2O)_m$ with $n \approx m$). Also called sugars. An example is glucose. Important source of chemical energy, 4

causal network A network in which the parents of a variable are interpreted as the immediate causes of the variable, 51

causal Markov assumption Given the value of a variable's immediate causes, its value of a variable is independent of all earlier causes, 68

cell The smallest structural and functional unit of an organism that is considered to be living, 9

cell membrane A cell's protective coat. The outer lining for a eukaryotic cell, 10

cell wall A rigid layer that surrounds the cell membrane and helps a cell maintain its shape, 10

cellular reproduction The process in which a cell forms a complete copy of itself, 9

centroid The average profile of the items within a cluster, 59

chemical bases The elements that make up DNA and RNA which include adenine (A), guanine (G), cytosine (C), thymine (T), and uracil (U), 4

chemical equilibrium A system in which all forward and reverse reactions are in balance resulting in no net change in the amounts or concentrations of any species, 104

chemical Langevin equation A chemical reaction model that is composed of both a deterministic component that grows linearly with respect to the propensity functions and a stochastic component that grows proportionally to the square root of the propensity functions, 118

chemical master equation (CME) A formal representation that describes the time evolution of the probabilities for the states in an SCK model, 103

chemical reactions The process of converting reactants into products, xxiv

chemical species Atoms or molecules involved in chemical reactions, xxiv

ChIP-on-chip This stands for chromatin immunoprecipitation-on-chip where the second chip refers to the use of a microarray chip in a later stage of processing. ChIP-on-chip is an experimental technique that is used to find protein-DNA interactions, 56

chloroplasts Organelles that generate energy in plant cells from sunlight using photosynthesis, 12

chromosome A DNA sequence containing genetic material for an organism. While prokaryotes have one chromosome which is circular, eukaryotes may have multiple chromosomes, 8

CI protein The phage λ repressor protein which is used to maintain the lysogeny pathway, 19

CII protein The phage λ protein that activates the P_{RE} promoter to initiate initial production of the CI protein, 29

CIII protein The phage λ protein that is responsible for protecting the CII protein, 29

circadian rhythms An approximately 24-hour cycle that helps regulate life's processes, 190

circularization The process in which a linear DNA strand forms a circular DNA strand when the "sticky ends" join, 29

cis-acting When a transcription factor affects adjacent genes, 15

classical chemical kinetics (CCK) A modeling method in which each chemical species concentration is tracked, and the changes in these concentrations are represented using ODEs, 85

cluster analysis The process of assigning objects into groups known as clusters in which objects within a cluster are more related in some way to objects within different clusters, 59

coding sequences DNA sequences that specify the order of the amino acids in a protein, 13

coding strand The strand of the DNA that has the code for a gene being transcribed. Its sequence is complementary to the template strand, 13

codon A sequence of three nucleotides which are associated to a particular amino acid by the genetic code, 15

codon Three bases which code for one of the 20 amino acids, 5

cohesive ends Single stranded ends to a linear DNA molecule which are complementary allowing them to bind to form a circular strand of DNA. Also known as sticky ends, 29

combinational logic gates Gates in which their output is strictly determined by their current input, 189

common cause A random variable that is a parent of two other random variables, 62

competitive enzymatic reactions A set of reactions representing a situation where multiple substrates require the same enzyme to produce their corresponding product, 136

complementary base pairs The notion that A always pairs with T (U) and G always pairs with C in DNA (RNA), 5

complementary DNA (cDNA) A single stranded DNA molecule that has the complementary sequence for a sequence of interest, 53

computational biology See bioinformatics, xxiii

conditional independence Two variables A and B are conditionally independent given a third variable C when the probability of the value of A is independent of the value of B given knowledge about the value of C, 62

consensus sequence When comparing several aligned DNA sequences, it is the sequence constructed from the most commonly found entries, 27

continuous-time Markov chain (CTMC) A Markov chain in which states can change at arbitrary points in time, 178

convergence The point in which the state distribution is not changing during an iterative method, 176

cooperativity A situation that occurs when there are multiple operator sites to which transcription factors can bind and one molecule binding assists other molecules to bind. For example, CI molecules can bind to the O_R operators sites in cooperative fashion, 24

cos site The location of the cohesive ends on the phage λ genome, 29

covalent bonds A bond formed by a shared electron in the valence band, 2

cristae Inward folds in the inner membrane of a mitochondria, 12

critical intervals The range of values between two critical thresholds, 164

critical threshold The value where a sigmoid or hill function transitions, 163

Cro protein A protein from phage λ which initiates the lysis pathway, 19

cytoplasm The large fluid-filled space inside the cell, 10

cytosine The chemical base "C" found in DNA and RNA, 5

cytoskeleton The cell's scaffold which organizes and maintains the cell's shape, 11

deoxyribose The sugar molecule found in the backbone of DNA, 4

dependency graph A data structure used by the next reaction method that is used to indicate how a reaction affects the propensities of other reactions, 112

dependent The condition where knowledge of the value of one random variable provides information about another random variable, 62

digestive enzymes Proteins that speed up biochemical processes, 12

digital circuit A circuit having discrete valued states, 163

dimer A molecule that consists of two like atoms or molecules. For example, CI can appear in either monomer or dimer form (denoted CI_2), 2

direct methods Methods of finding the steady-state distribution for a Markov chain which solve the system of equations given by $\pi = \pi P$, 176

direct method A version of Gillespie's SSA which randomly selects a reaction time followed by which reaction to execute, 111

discrete-time Markov chain (DTMC) A Markov chain in which states are only observed at discrete time points, 174

DNA A double-stranded sequence of nucleic acids connected by a sugar-phosphate backbone where the sugar is deoxyribose, 4

DNA folding Bending of a DNA molecule that can affect the transcription rate of a gene, 13

double-stranded Two complimentary strands of nucleic acids that are composed together to form DNA, 5

dual-rail A signal encoding which uses two signals to encode one value. One signal is high to indicate that the value is high while the other signal is used to indicate that the value is low. When both signals are low or both are high, the value of the signal is undefined, 202

dynamic Bayesian networks (DBN) A type of Bayesian network where each variable is given a time value between 1 and T. In other words, it is essentially a cyclic graph that has been unrolled T times, 68

E. coli Short for Escherichia coli which is a gram negative bacterium that is often found in the intestine of warm-blooded animals. *E. coli* is commonly used as a model organism in microbiology and as a host in synthetic biology, 17

early proteins Enzymes used by a virus to replicate nucleic acids, 17

electronic design automation (EDA) Tools for the automated design of electronic circuits and systems, 211

elemental reaction A reaction that can be considered a distinct physical event that happens nearly instantaneously, 104

elongation The second step of translation in which the ribosome moves along the mRNA transcript to construct the amino acid chain, 15

embedded Markov chain A DTMC used to find the steady-state distribution of a CTMC, 179

endocytosis The uptake of external materials by a cell, 11

endoplasmic reticulum The transport network within a cell which is used by molecules that are either targeted for modifications or for specific destinations, 12

entropy The amount of disorder in the universe, 2

enzymatic reactions Reactions that transform a substrate into a product catalyzed by an enzyme, 133

enzyme A reactant that accelerates a reaction without being consumed by the reaction. Also known as a catalyst, 4

equal data level assignment A level assignment which attempts to divide the data as equally as possible between the bins, 57

equal spacing level assignment An assignment of levels in which the levels divide the range of data values equally, 57

equilibrium constant The ratio of the forward over the reverse rates of a chemical reaction, 3

equilibrium point A state in which the rate of change of all variables is zero, 92

equivalence class A partition of a set of objects into groups where each item within a group is equivalent to all other items within the group, 63

equivalent Two items are the same, 63

ergodic Markov chain A Markov chain which is positive-recurrent and aperiodic, 175

eukaryotes Organisms that include a nucleus. They are usually multicellular such as fungi, mammals, birds, fish, invertebrates, mushrooms, and plants. They also include some complex single-celled organisms, 11

Euler's method A numerical solution to the initial value problem that assumes that rates of change remain constant during some time step, 88

exocytosis The process in which intracellular materials are released into a cell's environment, 17

exons Coding portions of a gene found in eukaryotes. They are broken up by introns that do not code for a protein, 9

explicit ODE method A numerical simulation method which calculates rates of change using only current state information, 88

factors Name used by Gregor Mendel for genes, 8

feedback Using an output signal as an input signal to create state, 198

first reaction method A version of Gillespie's SSA which determines the next reaction time for each reaction and executes the reaction with the earliest next reaction time, 111

fixed point See equilibrium point, 92

flagella An appendage to a cell which is used for locomotion, 10

flow The current rate of change of some variable, 92

forward Euler method A version of Euler's method which calculates the rate of change at each time point using only the current state information, 88

forward rate constant The rate constant associated with the kinetic law for the forward reaction of a reversible reaction, 86

fourth-order Runge-Kutta method A numerical simulation method for the initial value problem which determines the rates of change considering four points along a time step, 90

free energy The amount of energy associated with a particular operator site configuration. The more negative the value, the more likely the configuration, 145

gas constant 1.987 calories per mole. Usually expressed as R, 3

gene A sequence of nucleotides connected by their phosphate molecules that encode the instructions to construct a protein, 4

generalized logical models Logical models proposed by Thomas and Thieffry in which behavior is asynchronous and species are modeled by n-ary variables, 185

genetic circuits See genetic regulatory networks, xxiv

genetic code The code used to translate DNA sequences into amino acid sequences. Each three bases forms a codon that codes for one of the 20 amino acids, 5

genetic design automation (GDA) Tools for the automated design of genetic circuits and systems, 211

genetic engineering The process of directly manipulating an organisms genes, 187

genetic regulatory network A collections of genes, DNA binding sites, and proteins that interact to control the rate of transcription of the genes in the network, xxiv

genome All the genetic material within the chromosomes for an organism, 8

genome-wide location analysis See ChIP-on-chip, 56

Gibb's free energy The amount of total energy that is required for a reaction. A reaction can only occur spontaneously if this value is negative, 3

Gillespie's SSA See stochastic simulation algorithm, 107

glucose A carbohydrate with the chemical formula $C_6H_{12}O_6$, 4

glycoproteins A molecule that consists of both a carbohydrate and a protein. They can collect on a cell membrane to form an exit site for a virus, 17

Golgi apparatus Processes, packages, and transports proteins to be exported from the cell, 12

graphical user interface (GUI) A computer interface that allows the user to interact with the computer using icons and pictures rather than just text, xxiv

green fluorescent protein (GFP) A protein which comes from a jellyfish that fluoresces green when exposed to blue light, 52

guanine The chemical base "G" found in DNA and RNA, 5

guard A logical function composed of a conjunction of literals where each literal checks the value of a logical variable, 170

guarded commands A representation method for a stochastic FSM in which each guarded command includes a guard, a transition rate, and an assignment. The behavior is defined such that when the guard is true, the assignment can take place at the specified transition rate, 170

hidden common cause A random variable that is not measured but is actually the common cause for two other random variables, 62

hierarchical clustering An algorithm for cluster analysis which combines clusters two at a time which are the smallest distance apart. The result is a hierarchical tree representing the clusters, 59

Hill function A function of the form $1/(1 + Kx^n)$ or $Kx^n/(1 + Kx^n)$. They can be used to represent the effect of transcription factors on the rates of transcription, 138

homogeneous Markov process A Markov process which does not depend on the current time t, 174

hybrid Petri net (HPN) A Petri net that includes constructs to model both discrete and continuous states, 167

hydrogen bonds A weak bond between a hydrogen atom bonded to an O or N atom and an another atom containing an unshared electron pair that differ in how the electrons are shared between the atoms, 2

hydrolysis The splitting of a molecule into two or more smaller molecules by adding water, 3

hydrophilic Water-loving, 4

hydrophobic Water-fearing, 4

immediate causes The variables which directly determine a given variable's value, 68

implicit ODE method A numerical simulation method which determines the rate of change by solving a set of algebraic equations, 88

in silico Analysis on a computer, xxiv

inclusions Reserve deposits of a variety of different substances within a prokaryotic cell. These deposits may be membrane-bound, 10

independent The condition where knowledge of the value of one random variable does not give any information about another random variable, 62

indexed priority queue A data structure used by the next reaction method that organizes reactions by their reaction times making it easy to find the reaction with the smallest next reaction time, 112

induction event An event such as DNA damage which when detected causes a transition from the lysogeny pathway to the lysis pathway, 26

influence vector A representation which indicates which species influence a given gene, 69

influences The transcription factors that change the rate of transcription of a gene, 69

initial marking The number of tokens that are found initially in each place in a Petri net, 119

initial value problem The problem of determining the time evolution of the species concentrations given a set of ODEs and some initial condition, 88

initiation The first step of translation in which the ribosome attaches to the Shine-Delgarno sequence upstream from the start codon, 15

input function A function that indicates how many tokens in a Petri net are consumed from a place p when a transition t fires, 119

Int protein The protein responsible for integrating the phage λ genome within the host's chromosome, 29

integration The process in which a phage's genetic material is inserted into the host cell's genome, 34

interventions Changes in a network such as gene mutations, 69

introns DNA sequences found between exons which do not code for a protein, 9

inverter A gate which inverts the logical value of its input. In other words, its output is 1 when its input is 0, and it is 0 when its input is 1, 189

ionic bonds The force that holds ions together in a crystal (also known as an electrostatic bond), 2

irreducible DTMC A DTMC in which every state can be reached by every other state, 175

irrelevant node elimination A reaction-based abstraction that uses reachability analysis to detect nodes that do not influence the species of interest, 132

irreversible A reaction in which the forward reaction is so strongly thermodynamically favored over the reverse reaction that it is essentially impossible for the reverse reaction to occur, 86

isoelectric point The pH where a protein has no net change, 54

isopropyl β-D-thiogalatopyranoside (IPTG) A small molecule chemical inducer which binds to the protein LacI preventing it from being able to act as a transcription factor, 195

iterative methods Methods of finding the steady-state distribution for a Markov chain which apply the recursive formula $\pi(n) = \pi(n-1)P$, 176

joint probability distribution The probability distribution for multiple random variables, 62

jump Markov process A Markov process in which state updates occur in discrete amounts, 106

junk DNA Locations in a DNA sequence that have no known function, 9

K-means An algorithm for cluster analysis which partitions N genes into K clusters, 59

kinetic rate law A mathematical formula that governs the dynamics of a reaction, xxiv

labeled hybrid Petri net (LHPN) A hybrid Petri net that uses auxiliary variables to model continuous values. These values are checked and updated using labels on the transitions, 167

lambda repressor See CI, 19

large subunit Part of the ribosome with two sites to which tRNAs bind to construct a chain of amino acids, 15

late proteins Proteins used by viruses to construct the virus capsid, 17

law of mass action The speed of a chemical reaction is proportional to the amount of the reacting species, 2

laws of thermodynamics The 1st law states that energy is conserved. The 2nd law states that entropy increases. The 3rd law states that all processes stop at a temperature of absolute zero, 2

leaky gates Gates in which their logic function only represents their most likely behaviors, 168

leap condition The requirement that τ in the tau-leaping method must be chosen to be small enough such that no propensity function changes by a significant amount, 116

levels The values that are used to separate bins, 57

library protein A protein that is being checked to see if it interacts with the bait, 56

limiting distribution The probability vector for all states in a Markov chain that is reached in the limit as time goes to infinity, 175

lipid bi-layers Material formed when hydrophobic portions of two lipids bind to form a membrane, 4

lipids Made up of mostly of carbon and hydrogen atoms. Primary use is to form membranes. Examples of simple lipids include fats, oils, and waxes, 4

logical encoding Representing a continuous value using a mapping to a discrete number of logic levels, 161

lyse To break open the cell so that newly constructed viruses can exit, 17

lysis The process in which a virus opens the cell wall in order to exit the cell, 17

lysogens Cells with phage λ DNA integrated within their own DNA, 26

lysogeny The process in which a virus replicates its genetic material by integrating its DNA within its host's cell DNA and thus having it replicated through normal cell division, 17

lysosomes The digestive system of a cell. They are spherical, membrane bound, and made in the Golgi apparatus, 12

lysozyme An enzyme used to break down the cell wall, 17

lytic proteins Proteins produced by a virus to open the cell wall for exit, 17

macromolecules Large complex molecules such as carbohydrates, nucleic acids, and proteins, 4

major groove The larger gap in the DNA helix, 27

majority gate A combinational logic gate in which its output goes high when the majority of its inputs are high, 205

marginal likelihood In Bayesian analysis, this is a function over statistical model parameters that has been integrated over these parameters, 64

marker A sequence of DNA used to identify people, 9

marking The state of a Petri net which indicates how many tokens are currently found in each place, 119

Markov chain A Markov process that has a discrete state space, 174

Markov chain analysis Analysis methods which consider systems with the Markov property (i.e., the next state is only dependent on the current state), 174

Markov process A process where the next state is only dependent on the present state and not the past history, 106

Markov property The next state in a Markov process only depends on the current state, 174

Markov assumption Each variable X_i is independent of its non-descendants given its parents, 62

Markov relation There is Markov relation between genes X and Y when Y is a member of the minimal set of variables that when controlled for makes X independent of all other variables outside this set, 66

mass spectrometry An experimental technique in which a peptide is crystallized on a matrix, ionized with a laser, and the resulting ions are shot towards a detector. The higher the mass of an ion, the farther it travels, so the number of ions collected from a variety of collectors provides the molecular weights of the components of the peptide, 54

membranes Material used to separate cells from one another and create compartments within cells as well as having other functions, 4

memoryless The probability of a future state depends only on the present state, and it does not depend on any past states or the amount of time in which it has already spent in the current state., 174

messenger RNA (mRNA) RNA created from genes that code for proteins during transcription. This molecule is used to carry the genetic information encoded in the DNA to the ribosome as a template to synthesize the protein encoded by the gene, 5

metabolic network The enzymatic processes within a cell that transform food into energy, xxiv

methyl group A hydrophobic compound with the chemical formula -CH3, 15

methylation A chemical modification of DNA in which a methyl group is added to it, 15

Michaelis-Menten constant A constant formed from the rates in an enzymatic reaction used when performing a steady state approximation, 133

Michaelis-Menten equation The simplified form of an enzymatic reaction resulting from a quasi-steady-state approximation, 135

microarray A chip made of glass, plastic, or silicon to which an array of thousands or tens of thousands of single-stranded cDNA probes are attached, 53

microfilaments Long thin fibers within the cytoskeleton, 11

microtubules Hollow cylinders within the cytoskeleton, 11

midpoint method See second-order Runge-Kutta, 89

mitochondria Organelles that generate energy, 12

mitochondrial genome A genome found in the mitochondria. It is a circular DNA molecule that is inherited only from the mother. Although very small, its genes code for some important proteins, 12

mitochondrial theory of aging Theory that suggests that mutations in mitochondria may drive the aging process, 12

modifier A chemical species that is used in a chemical reaction but is neither created nor destroyed by that reaction. Also known as an enzyme, xxiv

modifier constant propagation A reaction-based abstraction that removes modifiers that are constant (i.e., species that are neither produced nor destroyed), 144

molecular chaperons Proteins used by viruses to assist in assembly, 17

molecular species Atoms or molecules involved in chemical reactions, 2

molecules Collections of atoms bonded together, 1

monomer A molecule that consists of just one copy of some molecule. See also dimer, 19

monomolecular reaction channel A reaction with only one reactant, 105

morphogens A substance suggested by Turing that chemically diffuses between cells to indicate to each cell what type of tissue to develop into, 98

Muller C-element A sequential logic gate which sets its output high when all of its inputs are high, and it sets it low when all of its inputs are low. Otherwise, it retains its previous state, 202

multicellular An organism composed of multiple cells, 9

multiple product splitization A reaction-based abstraction that splits an irreversible reaction with multiple products into multiple irreversible reactions each with a single product, 170

multiple reactant / multiple product reaction splitization A reaction-based abstraction that splits an irreversible reaction with multiple reactants and multiple products into an irreversible reaction with multiple reactants and no products and an irreversible reaction with no reactants and multiple products, 170

multiple reactant splitization A reaction-based abstraction that splits an irreversible reaction with multiple reactants into multiple irreversible reactions each with a single reactant and the other reactants as modifiers, 170

multiplicity of infection The number of viruses that have infected a cell, 33

mutations Small random changes such as in a DNA sequence, 6

N protein The phage λ protein, also known as the anti-terminator, which is responsible for binding to RNAP at the nut sites to allow the RNAP to pass over terminator sites, 29

n-ary encoding A generalization of binary encoding to more than just two discrete values, 163

NAND gate A combinational logic gate that has its output go low only when all of its inputs are high, 191

nascent Newly formed, 13

negative feedback Feedback that is used to inhibit further change, 198

next reaction method A variation on the first reaction method proposed by Gibson and Bruck which uses clever data structures to avoid recalculating propensities and next reaction times when it is unnecessary, 112

non-critical reactions A reaction that can be fired a given small number of times without causing one of its reactant species counts to become negative, 116

NOR gate A combinational logic gate that has its output go high only when all of its inputs are low, 191

nuclear membrane Spheroid shaped membrane that encloses the nucleus in a eukaryote, 12

nucleic acids The macromolecules that store information within living organisms. It is composed of a base bound to a sugar molecule and a phosphate molecule, 4

nucleotide One of the chemical bases found in DNA or RNA, 4

nucleus A membrane enclosed compartment in a eukaryote which contains the cell's genome. It is the location where DNA replication and transcription occurs, 11

null-recurrent state A state in a DTMC in which its mean time to be revisited is infinite, 175

nullclines A line in which the flow of a variable is zero, 95

nut site Location on a DNA sequence where the anti-terminator N can be bound to an RNAP molecule that is moving along the sequence, 30

O protein The phage λ protein which with the P protein assist with replication of the λ genome, 29

O_L operator The left operator site from phage λ which is responsible for controlling the transcription of the genes for N, CIII, Xis, Int, and Sib, 26

O_R operator The right operator in phage λ which is responsible for controlling the production of CI and Cro. It is also known as the genetic switch, since only one of these proteins is typically produced at a time, 19

obligate intracellular parasites Another name for viruses since they must utilize the machinery and metabolism of their host cell to reproduce, 16

one-dimensional ODE model An ODE model that includes only one variable, 92

open complex formation The process in which RNAP after binding to a promoter opens the DNA molecule in order to begin transcription, 44

open reading frames Sequences of 100 bases without a stop codon, 8

operator site reduction A reaction-based abstraction method that is used to simplify reactions involved in the binding of transcription factors to operator sites as well as those involved in open complex formation and production of a protein via transcription and translation, 136

operator sites Portions of the DNA sequence near the promoter where transcription factors bind, 13

OR gate A combinational gate that has its output go high when any of its inputs go high, 193

order relation There is an ordering relation between genes X and Y if X is an ancestor of Y, 66

ordinary differential equation (ODE) A function that includes only one independent variable and one or more of its derivatives, 85

organelles Membrane-bounded compartments in which specific metabolic activities are performed, 10

oscillator A circuit in which each signal alternates between a low and high value in a periodic fashion, 190

output function A function that indicates how many tokens in a Petri net are produced into a place p when a transition t fires, 119

P protein The phage λ protein which with the O protein assist with replication of the λ genome, 29

P site The location on the large subunit of the ribosome where the tRNA moves to after binding to the A site, and in this location it binds its amino acid to the growing chain, 15

P_I The promoter from phage *lambda* which is responsible for initiating transcription of the *int* gene, 30

P$_L$ promoter The left promoter from phage λ which is responsible for initiating transcription of the N gene as well as the *cIII*, *xis*, and *int* genes and the *sib* region, 30

P$_R$ promoter The right promoter from phage λ which is responsible for initiating transcription of the *cro* gene as well as the *cII*, *O*, *P*, and *Q* genes. It overlaps the O_R1 and part of the O_R2 operator sites, 19

P$_{antiq}$ promoter The promoter from phage λ that produces reverse transcripts for the gene Q which can bind with mRNA for the forward transcripts for Q preventing synthesis of Q, 30

P$_{R'}$ promoter The promoter from phage λ which is responsible for initiating transcription of genes needed for the lysis pathway, 30

P$_{RE}$ promoter The promoter from phage λ that initiates transcription of the *CI* gene when activated by the CII protein, 30

P$_{RM}$ promoter The repressor maintenance promoter from phage λ which is responsible for initiating transcription of the *cI* gene. It overlaps the O_R3 and part of the O_R2 operator sites, 19

parent When an arc in a Bayesian network is directed from X to Y, X is a parent of Y, 62

partial bin assignment A bin assignment where some species are not assigned a bin, 57

partial differential equation A function of several independent variables and their partial derivatives. They can be used to model spatial effects because they include rates both with respect to time and space, 85

partially directed acyclic graph A graph without cycles in which some arcs in the graphs are assigned a direction, 63

penetration The second stage of virus replication in which the virus injects its genetic material (with or without the capsid) inside the host, 16

peptide Part of a protein, 54

periodic Markov chain A Markov chain with a period greater than 1, 177

periodic state A state which upon leaving can only be returned to after a number of transitions that is a multiple of a period p where $p > 1$, 175

periodicity The period of a DTMC, 176

peroxisomes Part of the digestive system of a cell. They are similar to lysosomes, but they can self-replicate, 12

phage λ A bacteriophage that infects *E. coli* and can replicate its genetic material either through a lysis or lysogeny pathway. Phage λ is often used as a model organism in microbiology, xxiv

phase space The set of all possible states for a system described by a set of ODEs, 92

phosphate A molecule which is composed of phosphorous and oxygen, 4

photosynthesis The process that converts sun's light energy into ATP, 12

piecewise linear differential equation (PLDE) A differential equation which defines the rate of change using a linear equation for each domain within the state space, 165

piecewise model A model in which the state space is divided into domains, and behavior is defined independently for each domain, 165

pili Protein appendages which are attached to the cell surface and used for adhesion, 10

pitchfork bifurcation A behavior of a one-dimensional ODE model where as a parameter changes value there comes a point where the one stable equilibrium point becomes unstable and two new stable equilibrium points appear, 94

places The nodes of a Petri net that contain tokens that indicate the current state such as the number of molecules of a chemical species. They are represented with circles in a Petri net diagram, 119

Poisson random variable A random variable that can by used, for example, to return the number of events in a time interval, 116

polymerase chain reaction (PCR) An experimental technique for making many copies of a DNA sequence, 187

PoPS gates Synthetic genetic gates that use RNA polymerases that move across a DNA strand per second (i.e., PoPS) to indicate whether a gate is on or off, 198

positive feedback Feedback that is used to enhance further change, 198

positive recurrent state A state in a DTMC in which its mean time to be revisited is finite, 175

post-transcriptional modifications Changes made to an mRNA after transcription such as the removal of introns, 13

post-translational modifications Changes made to an amino acid chain recently produced by the translation process potentially changing the function of the resulting protein, 15

primary structure The sequence of amino acids that make up a protein, 6

priors probability distributions that represent knowledge about an unknown before any data has been considered, 64

production-passage-time approximation (PPTA) An enzymatic approximation that removes the unproductive reaction which breaks up the enzyme-substrate complex, 133

products The chemical species produced by a reaction, xxiv

prokaryotes Unicellular organisms that lack a nuclear membrane. Bacteria are an example, 10

prokaryotic cell membrane The protective coat that lines a prokaryotic cell, 10

promoter sequence A unidirectional sequence that is found on one strand of the DNA in front of a gene. It is recognized by RNAP as the location where transcription should be initiated, 13

propensity function The product of the specific probability rate constant with the number of possible combinations of reactant molecules. When multiplied by an infinitesimal time interval, the propensity function yields the probability of the corresponding reaction occurring within this time interval, 104

proteases Enzymes that degrade proteins, 32

protein folding The process under which a protein forms it shape after translation, 6

protein network A sequence of biochemical reactions used by a cell to convert one type of signal or stimulus into another type. This chain of reactions is also known as a signal transduction network or signaling cascade. These networks are used to respond to a variety of stimuli from the environment ranging from temperature and light changes to the presence of chemicals, xxiv

proteins The basic building blocks of nearly all molecular machinery of an organism. Made from long chains of amino acids in which their sequence is encoded in genes, 4

purines The bases adenine (A) and guanine (G), 5

pyrimidines The bases cytosine (C) and thymine (T), 5

Q protein The phage λ protein which binds to RNAP at the Qut site to activate transcription of the lysis, head, and tail genes, 29

qualitative logical model A purely logical model in that no rates, probabilities, or other quantitative parameters are provided, 180

qualitative differential equations A form of differential equations in which no explicit values are used for parameters or states, 185

qualitative ODE analysis Analysis of an ODE model to find the complete phase space for any initial condition as well as how it changes with parameter variations, 92

quasi-steady-state approximation A reaction-based abstraction for enzymatic reactions in which the amount of complex is assumed to be in steady-state allowing the model to be reduced to just one reaction that converts substrate directly into product, 135

quaternary structure The arrangement of proteins that are composed of multiple amino acid chains, 6

Qut site Location on the DNA sequence where the anti-terminator protein Q can bind to an RNAP molecule moving along the DNA, 33

rapid equilibrium approximation A further reduction beyond the quasi-steady-state approximation in which the rate for an enzymatic model is simplified by assuming that the production of product is the rate limiting step, 135

rate constant A value that indicates how likely or how fast a reaction typically occurs, 2

reactants The chemical species consumed by a reaction, xxiv

reaction rate equations A set of ODEs which describe the rates of change of concentration for each chemical species in a chemical reaction model, 85

reaction splitization A reaction-based abstraction method that splits a single reaction into multiple reactions, 170

reaction-based abstraction Abstraction methods that convert a chemical reaction model into a simpler one with potentially less species and/or reactions, 131

reaction-diffusion equations An ODE model that includes spatial effects by splitting up the space into regions and introducing variables representing the amount of each variable within each region, 97

recombinant DNA An experimental technique for constructing artificial DNA sequences through combinations, 187

recombination Large-scale rearrangements of the DNA molecule, 9

recurrent state A state in which a DTMC is guaranteed to return at some point in the future, 175

regulatory sequences Locations in a DNA sequence which mark the start and end of genes as well as those that are used to switch genes on and off, 8

regulatory domains A region in the space of potential species concentrations which is defined by the critical thresholds, 164

relative error level The desired maximum acceptable error relative to the actual values of the variables, 90

release The fifth and final stage of virus replication in which the new viruses are released either by lysing the cell wall or through exocytosis, 16

repetitive DNA Short sequences which are often repeated 100s of times in a DNA sequence, 9

replication The third stage of virus replication in which the virus takes over the cell to produce the materials needed to construct new viruses, 16

reporter gene A gene that produces a protein such as GFP that can be observed externally to indicate actions within a cell, 52

repress The binding of proteins to an operator site to turn off transcription of a gene, 13

repressilator A synthetic genetic circuit oscillator designed by Elowitz and Leibler, 190

repressor A transcription factor that precludes transcription of a gene usually by blocking RNAP from binding to the prompter, 15

retroviruses Viruses that use RNA as their genetic material, 16

reverse rate constant The rate constant associated with the kinetic law for the reverse reaction of a reversible reaction, 86

reverse transcriptase An enzyme that catalyzes the formation of a double-stranded DNA molecule from a single-stranded RNA template, 17

reversible reaction A reaction that can occur in both the forward and reverse directions, 3

ribose The sugar molecule found in the backbone of RNA, 4

ribosomal RNA (rRNA) RNA that forms the ribosome, 5

ribosome The protein assembly machinery that turns mRNA templates into proteins during translation, 5

ribosome binding sites A DNA sequence that promotes ribosome binding to initiate translation, 188

RNA A single-stranded sequence of nucleic acids connected by a sugar-phosphate backbone where the sugar is ribose, 4

RNA replicase An enzyme that catalyzes RNA replication, 17

RNA polymerase (RNAP) An enzyme that initiates DNA transcription, 13

saddle-node bifurcation A behavior of a one-dimensional ODE model in which as a parameter value changes there comes a point in which the model changes from one with no equilibrium points into one with one stable and one unstable equilibrium point, 93

second-order Runge-Kutta method A numerical simulation method for the initial value problem which determines the rates of change considering two points along a time step, 89

secondary structure The patterns formed by the amino acids that are near to each other in a protein. Examples include α-helices and β-pleated sheets, 6

sense strand See coding strand, 13

Shine-Delgarno sequence The DNA sequence (AGGAGGU) which is recognized and attached to by a ribosome to initiate translation, 15

sigmoid functions A function that has a low region and a high region separated by a steep transition region. The curve representing this function has an S-like shape, 138

similar reaction combination A reaction-based abstraction that combines reactions that have the same reactants, modifiers, and products, 144

single-rail The normal logical encoding of a signal in which its truth value is directly given by the value of the signal, 202

single-step transition probabilities The probability of moving from one state to another in one discrete step of a Markov chain, 174

sliding motion The possibility of sliding when a trajectory encounters a black or white boundary, 167

small subunit Part of the ribosome which attaches to the mRNA to initiate translation, 15

small nuclear RNAs (snRNAs) Small RNA molecules that are found in the nucleus of eukaryotic cells and are involved in a number of important processes, 5

sparse candidate algorithm A method to explore potential Bayesian networks in which a small number of potential parents are considered for each gene using local statistics, 64

spatial effects The effects that the location of the molecules within a cell have on the rates and probabilities of reactions, 86

spatial Gillespie method A Gillespie SSA method that accounts for spatial effects by dividing the space into discrete subvolumes, adding reactions to move species between the subvolumes, and tracking the number of species within each subvolume, 124

specific probability rate constant When multiplied by an infinitesimal time interval, this constant represents the probability that a randomly chosen combination of reactant molecules react within this time interval. It is related to the reaction rate constant, 104

speed-independent circuit A circuit that operates correctly regardless of the delays of any of its gates assuming that wire delays are negligible, 205

splice junctions Locations in a DNA sequence that separate exons from introns, 8

stability The determination if an equilibrium point is stable or unstable, 92

stable An equilibrium point in which the flow near the point is directed into the point, 92

standard free energy See Gibb's free energy, 3

standards The engineering principle of using repositories of reusable parts that are created such that they can be easily composed, 187

start codon The first codon in an mRNA sequence coding for a protein which indicates the starting point for translation. It is usually AUG (ATG on the DNA sequence) which codes for methionine, 8

state-change vector A vector that represents how a chemical reaction changes the amounts of each species, 104

states The values of all the variables in a model. For an ODE model of a set of chemical reactions, these values would include the concentration of each species and its current rate of change, 92

statistical confidence A measure that indicates the likelihood that a feature of a network is a true feature, 66

statistical thermodynamical model A model proposed by Shea and Ackers that removes rapid reactions involved in binding and unbinding of transcription factors to operator sites, and it assumes that the occupancy of operator sites can be determined by equilibrium statistical thermodynamic probabilities, 145

steady state A point in which the forward and reverse reactions have equal rates of reaction, 3

steady-state distribution The limiting distribution for an ergodic Markov chain, 175

steady-state assumption An assumption in which the rate of change of some species is assumed to be zero, 133

step doubling A technique that determines the necessary step size by comparing the results obtained after both one time step and one-half time step, 90

step size The amount of time advanced between two iterations of a numerical solution of the initial value problem for a set of ODEs, 88

stiff equations A set of ODEs that require a very small relative time step to produce accurate results, 89

stochastic chemical kinetics (SCK) A modeling method in which the amount of each chemical species is tracked, and the changes in these amounts are updated stochastically, 103

stochastic finite-state machine (FSM) A stochastic model that includes a finite number of discrete (logical) valued states and transitions between them that are annotated with the likelihood of making the transition, 170

stochastic Petri net (SPN) An alternative modeling formalism for SCK models that includes places that can represent species amounts, transitions that change species amounts, and a flow relation that represents how reactions change species amounts, 103

stochastic simulation A simulation method in which state updates are determined randomly, 107

stochastic simulation algorithm (SSA) A simulation algorithm that steps over time steps in which no reaction occurs. In other words, time does not evolve uniformly, but it instead jumps from the time of one reaction to the next, 107

stoichiometric matrix A two-dimensional array composed of the state change vectors, 104

stoichiometry The numbers of molecules created or destroyed by a reaction, 2

stoichiometry amplification A reaction-based abstraction method that increases the stoichiometry of all or part of the reactions while reducing their rates, 154

stop codons One of the three triplets of nucleotides that indicate that translation should terminate, 15

structural genes DNA sequences that code for proteins, 9

substrates Chemical species that are consumed in a reaction. Also known as reactants, 4

switch-level simulation A simulation method in which each signal value is assumed to take one of two binary values, 163

switching time The time it takes to change the value of a gate, 205

symbiotic Two or more organisms living together for their mutual benefit, 12

synthetic biology Genetic engineering which leverages the engineering principles of separating design issues from construction issues, using standard parts, and employing high-level abstractions, 187

synthetic genetic circuits Genetic circuits that do not occur naturally, but they have been constructed artificially for scientific, industrial, or medical use, 187

systems biology The study of the mechanisms underlying complex molecular processes as integrated into systems or pathways made up of many interacting genes and proteins, xxiii

systems biology markup language (SBML) An XML-based language for representing chemical reaction networks, xxiv

tau-leaping A stochastic simulation method proposed by Gillespie and his colleagues which allows many reactions to fire during each time step, τ, 115

template strand A strand of DNA used to construct a mirror image using the complimentary property of chemical bases, 5

template strand The strand of the DNA that is used during transcription. Its sequence is complementary to the coding strand, 13

termination The third and final step of translation in which the ribosome detects a stop codon and releases the mRNA template and newly formed protein, 15

terminator A DNA sequence that terminates transcription, 188

terminator switch A location in a DNA sequence which can terminate transcription. It is called a switch because RNAP that is bound to an anti-terminator protein is not stopped by a terminator switch, 30

ternary structure The arrangement of amino acids that are far apart in a protein, 6

thermal equilibrium A system which maintains a constant temperature, 104

thymine The chemical base "T" found in DNA, 5

time series experimental data Experimental data collected over a series of time points, 54

toggle switch A sequential genetic circuit which has its output set by one input and reset by another. This is more commonly known as a set-reset flip-flop, 201

training set A group of experimental data used to infer a Bayesian network, 63

trans-acting When a transcription factor affects distant genes, 15

transcription The process in which DNA is used as a template to create an mRNA copy that will be translated into a protein, 5

transcription factors Proteins that bind to operator sites to activate or repress transcription of a gene, 13

transcritical bifurcation A behavior of a one-dimensional ODE model with two equilibrium points, one stable and one unstable, in which as a parameter changes value there comes a point where the points cross and the stable point becomes unstable while the unstable point becomes stable, 94

transfer RNA (tRNA) RNA used during translation to shepherd amino acids to the ribosome that match a 3 base pair codon, 5

transient state A state in which there is a non-zero probability that the DTMC will at some point never return to that state, 175

transition probability matrix A matrix representation of the single-step transition probabilities for a homogeneous DTMC, 174

transition rate matrix A matrix representation for a CTMC in which each entry is the transition rate between two states, 178

transitions The nodes of a Petri net that represent the events that move tokens between places. They are represented with lines or boxes in a Petri net diagram, 119

translation The process in which an mRNA is used by a ribosome to construct a protein from the amino acid sequence encoded in the mRNA, 5

translational regulation Binding of repressor proteins to an mRNA molecule to inhibit its translation. This type of regulation is heavily utilized during embryonic development and cell differentiation, 15

transparent boundary A boundary between two domains in a piecewise differential equation model in which all trajectories enter one domain and leave the other domain through this boundary, 167

trimolecular reaction channel A reaction with three reactants, 105

two-dimensional gel electrophoresis An experimental procedure that separates proteins using an electric field in one dimension and their pH in the second dimension, 54

two-dimensional ODE model An ODE model that has two variables, 94

two-hybrid screening An experimental technique that is used to determine if two proteins interact, 55

unicellular An organism composed of a single cell, 9

universal gate A combinational gate from which any other combinational gate can be constructed. A NAND gate is an example of one, 191

unstable An equilibrium point in which the flow near the point is directed out of the point, 92

uracil The chemical base "U" found only in RNA, 5

vector A carrier of a virus, 16

vector fields A graphical technique used to visualize two-dimensional ODE models in which the axes are the two variables and each point is associated with a vector indicating the direction of flow, 95

viral capsid The protein package that surrounds the genetic material of a virus, 16

virion The name given to a virus before entering the host which means a package of genetic material, 16

viruses Genetic material, DNA or RNA, surrounded by a protein package. Since they do not have a metabolic system and rely upon the host cells that they infect in order to reproduce, they are not considered to be living, 16

weight function A function that assigns a rate or probability to a transition in a stochastic Petri net, 119

well-stirred An assumption that molecules are equally distributed within a cell. This assumption allows spatial effects to be neglected, 86

white boundary A boundary between two domains in a piecewise differential equation model in which all trajectories enter both domains from this boundary, 167

wild-type The form that an organism, gene, etc. takes normally in nature. For example, a wild-type gene is not mutated, 24

Xis protein The protein responsible for excising the phage λ genome out of the host's chromosome, 29

Index

λ repressor, 19
3' end, 5
5' end, 5

A site, 15
absolute error level, 90
abstraction, 187
acceptor site, 15
accuracy control parameter, 116
activate, 13
activated rate, 20
activating domain, 55
activation energy, 3
activator, 13
adaptive stepsize control, 90
adenine, 5
affinity, 20
agglomerative hierarchical clustering, 59
alternative splicing, 9
amino acids, 5
analog circuit, 163
AND gate, 193
animal viruses, 16
anti-codon site, 15
anti-terminator, 30
antisense strand, 13
aperiodic state, 175
assembly, 16
aTc, 195
atoms, 1
ATP, 3
attachment, 16
attachment site, 29
attp site, 29
automated construction, 187
automated sequencing, 187
autoregulation, 198
average phage input, 122

background knowledge, 70
backward Euler method, 88

bacteria, 9
bacteriophages, 16
bait, 56
basal rate, 20
base bin assignment, 72
Bayesian learning, 63
Bayesian networks, 51, 62
Bayesian scoring metric, 64
bifurcation diagram, 93
bifurcation point, 92
bimolecular reaction channel, 104
bin assignment, 57
binding domain, 55
bins, 57
BioBricks^{TM}, 188
bioinformatics, xxiii
biological databases, xxv
bistability, 198
black boundary, 167
bootstrap, 67

capsule, 10
carbohydrates, 4
causal Markov assumption, 68
causal network, 51, 68
cell membrane, 10, 11
cell wall, 10
cells, 1, 9
centroid, 59
chemical equilibrium, 104
chemical base, 4
chemical Langevin equation, 118
chemical master equation, 103, 107
chemical reactions, xxiv, 1
chemical species, xxiv, 2
ChIP-on-chip, 54, 56
chloroplasts, 12
chromosome, 8, 11
CI protein, 19, 29
CII protein, 29
CIII protein, 29

circadian rhythms, 190
circularization, 29
cis-acting, 15
classical chemical kinetics, 85
cluster analysis, xxiii, 51, 59
coding sequences, 13
coding strand, 13
codon, 5, 15
cohesive ends, 29
combinational logic gates, 189
common cause, 62
competitive enzymatic reactions, 136
complementary base pairs, 5
complementary DNA, 53
computational biology, xxiii
conditional independence, 62
consensus sequence, 27
continuous-time Markov chain, 178
convergence, 176
cooperativity, 24, 205
cos site, 29
covalent bonds, 2
cristae, 12
critical intervals, 164
critical threshold, 163
Cro protein, 19, 29
cytoplasm, 10, 11
cytosine, 5
cytoskeleton, 11

deoxyribose, 4
dependency graph, 112
dependency graph, 114
dependent, 62
digestive enzymes, 12
digital circuit, 163
dimer, 2, 19
dimerization reduction, 131
direct methods, 176
discrete-time Markov chain, 174
DNA, 4
DNA folding, 13
DNA microarrays, 51, 53
double-stranded, 5
dual-rail, 202
dynamic Bayesian networks, 68

early proteins, 17
electronic design automation, 211

elemental reaction, 104
elongation, 15
embedded Markov chain, 179
endocytosis, 11, 16
endoplasmic reticulum, 12
entropy, 2
enzymatic reactions, 133
enzymatic approximation, 131
enzyme, 4
equal data level assignment, 57
equal spacing level assignment, 57
equilibrium constant, 3
equilibrium point, 92
equivalence class, 63
equivalent, 63
ergodic Markov chain, 175
eukaryotes, 11
Euler's method, 88
exocytosis, 17
exons, 9
explicit ODE method, 88

factors, 8
feedback, 198
fixed point, 92
flagella, 10
flow, 92
forward Euler method, 88
forward rate constant, 86
fourth-order Runge-Kutta method, 90
free energy, 145

gas constant, 3
gene, 4, 8, 188
generalized logical models, 185
genetic engineering, 187
genetic circuits, xxiv, 13
genetic code, 5
genetic design automation, 211
genetic regulatory network, xxiv
genome, 8–10, 29
genome-wide location analysis, 54, 56
genomes, 1, 8
Gibb's free energy, 3
Gillespie's SSA, 107
glucose, 4
glycoproteins, 17
Golgi apparatus, 12
graphical user interfaces, xxiv

green fluorescent protein, 52, 189
guanine, 5
guard, 170
guarded commands, 170

hidden common cause, 62
Hill function, 138, 140, 161, 163
homogeneous Markov process, 174
hybrid Petri net, 167
hydrogen bonds, 2
hydrophilic, 4
hydrophobic, 4
hyrdrolysis, 3

immediate causes, 68
implicit ODE method, 88
in silico, xxiv, 85
inclusions, 10
independent, 62
index, 95
indexed priority queue, 112
induction, 34
induction event, 26
influence vector, 69
influences, 69
initial marking, 119
initial value problem, 88
initiation, 15
input function, 119
Int protein, 29
integration, 34
introns, 9
inverter, 189
ionic bonds, 2
IPTG, 195
irreducible DTMC, 175
irrelevant node elimination, 131, 132
irreversible, 86
isoelectric point, 54
iterative methods, 176

joint probability distribution, 62
jump Markov process, 106
junk DNA, 9

K-means, 59
kinetic rate law, xxiv

labeled hybrid Petri net, 167

large subunit, 15
late proteins, 17
law of mass action, 2, 85, 86, 133, 138,
    139, 141
laws of thermodynamics, 2
leaky gates, 168
leap condition, 116
levels, 57
library protein, 56
limiting distribution, 175
lipid bi-layers, 4
lipids, 4
logical encoding, 161, 163
lyse, 17
lysis, 17
lysogens, 26
lysogeny, 17
lysosomes, 12
lysozyme, 17
lytic proteins, 17

macromolecules, 1, 4
major groove, 27
majority gate, 205
marginal likelihood, 64
marker, 9
marking, 119
Markov assumption, 62
Markov chain, 174
Markov chain analysis, 161, 174
Markov process, 106, 174
Markov property, 174
Markov relation, 66
mass spectrometry, 54
membranes, 4
memoryless, 174
Mendel, Gregor, 8
messenger RNA, 5
metabolic networks, xxiv
methyl group, 15
methylation, 15
Michaelis-Menten constant, 133
Michealis-Menten equation, 135
microarrays, 53
microfilaments, 11
microtubules, 11
midpoint method, 89
mitochondria, 12
mitochondrial genome, 12

mitochondrial theory of aging, 12
modifier, xxiv, 4
modifier constant propagation, 144
molecular chaperons, 17
molecular species, 2
molecules, 1
monomer, 19
monomolecular reaction channel, 105
morphogens, 98
Muller C-element, 202
multicellular, 9
multiplicity of infection, 33, 122
mutations, 6, 9, 12, 24

N protein, 29
n-ary encoding, 163
NAND gate, 191
nascent, 13
negative feedback, 198
non-critical reactions, 116
NOR gate, 191
nuclear membrane, 12
nucleic acids, 4
nucleotide, 4
nucleus, 11
null-recurrent state, 175
nullclines, 202
nut site, 30

O protein, 29
O_L operator, 26
O_R operator, 19
obligate intracellular parasites, 16
open complex formation, 44
open reading frames, 8
operator site reduction, 131, 136
operator sites, 13, 26
OR gate, 193
order relation, 66
ordinary differential equations, 85
    one-dimensional model, 92
    two-dimensional model, 94
organelles, 10, 11
oscillator, 190
output function, 119

P protein, 29
P site, 15
P_I promoter, 30

P_L promoter, 30
P_{antiq} promoter, 30
$P_{R'}$ promoter, 30
P_{RE} promoter, 30
P_{RM} promoter, 19, 30
P_R promoter, 19, 30
parent, 62
partial bin assignment, 57
partial differential equations, 85, 97
partially directed acyclic graph, 63
penetration, 16
peptide, 54
periodic state, 175
periodic Markov chain, 177
periodicity, 176
peroxisomes, 12
Petri nets, 119
phage λ, xxiv, 17
phase space, 92
phosphate, 4
photosynthesis, 12
piecewise linear differential equation, 165
piecewise models, 161, 165
pili, 10
pitchfork bifurcation, 94
places, 119
Poisson random variable, 116
polymerase chain reaction, 187
PoPS gates, 187, 198
positive feedback, 198
positive-recurrent state, 175
post-transcriptional modifications, 13
post-translational modification, 15
PPTA, 133
primary structure, 6
priors, 64
products, xxiv, 2
prokaryotes, 10
prokaryotic cell membrane, 10
promoter sequence, 13
propensity function, 104
proteases, 32, 49
protein folding, 6
protein networks, xxiv
proteins, 1, 4, 5
purines, 5
pyrimidines, 5

Q protein, 29

qualitative differential equations, 185
qualitative logical models, 161, 180
qualitative ODE analysis, 92
quasi-steady-state approximation, 135
quaternary structure, 6
Qut site, 33

rapid equilibrium approximation, 135
rate constant, 2
reactants, xxiv, 2
reaction splitization, 170
reaction rate equations, 85, 88
reaction-based abstractions, 131
reaction-diffusion equations, 97
recombinant DNA, 187
recombination, 9
recurrent state, 175
regulatory domains, 164
regulatory sequences, 8, 9, 13, 188
relative error level, 90
release, 16
repetitive DNA, 9
replication, 12, 16
reporter gene, 52, 188
repress, 13
repressilator, 190
repressor, 15, 136
reproduce, 9
retroviruses, 16, 17
reverse rate constant, 86
reverse transcriptase, 17, 53
reversible reaction, 3
ribose, 4
ribosomal RNA, 5
ribosome, 5, 10, 12, 15
ribosome binding sites, 188
RNA, 4, 5
RNA polymerase, 13
RNA replicase, 17

saddle-node bifurcation, 93
SBML, xxiv, 211
second-order Runge-Kutta method, 89
secondary structure, 6
sense strand, 13
Shine-Delgarno sequence, 15
sigmoid functions, 138
similar reaction combination, 144
single-rail, 202

single-step transition probabilities, 174
sliding motion, 167
small nuclear RNAs, 5
small subunit, 15
sparse candidate algorithm, 64
spatial effects, 86
spatial Gillespie method, 124
specific probability rate constant, 104
speed-independent circuit, 205
splice junctions, 8
SSA, 107
stability, 92
stable, 89, 92
standard free energy, 3
standards, 187
start codon, 8, 15
state-change vector, 104
states, 92
statistical confidence, 66
statistical thermodynamical model, 145
statistical thermodynamics, 131
steady state, 3
steady-state assumption, 133
steady-state distribution, 175
step doubling, 90
step size, 88
stiff equations, 89
stochastic simulation, 107
stochastic chemical kinetics, 103
stochastic finite-state machines, 161, 170
stochastic Petri nets, 119
stochastic simulation algorithm, 107
    direct method, 111
        first reaction method, 111
        next reaction method, 112
stoichiometric matrix, 104
stoichiometry, 2
stoichiometry amplification, 131, 154
stop codon, 15
structural genes, 9
substrates, 4
switch-level simulation, 163
switching time, 205
symbiotic, 12
synthetic biology, 187
synthetic biology, 39
synthetic genetic circuits, 187
systems biology, xxiii
systems biology markup language, xxiv

tau-leaping, 115
template strand, 5, 13
termination, 15
terminator, 188
terminator switch, 30
ternary structure, 6
thermal equilibrium, 104
thymine, 5
time series experimental data, 54, 57
toggle switch, 201
training set, 63
trans-acting, 15
transcription, 5, 12, 13
transcription factors, 13
transcritical bifurcation, 94
transfer RNA, 5, 15
transient state, 175
transition probability matrix, 174
transition rate matrix, 178
transitions, 119
translation, 5, 12, 13, 15
translational regulation, 15
transparent boundary, 167
trimolecular reaction channel, 105
two-dimensional gel electrophoresis, 54
two-hybrid screening, 54, 55

unicellular, 9
universal gate, 191
unstable, 92
uracil, 5

vector, 16
vector fields, 95
viral capsid, 16
virion, 16
viruses, 1, 4, 16

weight function, 119
well-stirred, 86, 104
white boundary, 167
wild-type, 24

Xis protein, 29

Printed and bound by CPI Group (UK) Ltd, Croydon, CR0 4YY

21/10/2024

01777089-0010